Lecture Notes in Mathematics

Edited by A. Dold and B. Eckmann

1229

Ola Bratteli

Derivations, Dissipations and Group Actions on C*-algebras

Springer-Verlag

Berlin Heidelberg New York London Paris Tokyo

Author

Ola Bratteli
Institute of Mathematics, University of Trondheim,
N-7034 Trondheim-NTH, Norway

Mathematics Subject Classification (1980): 46L55, 22D25, 20M20, 34C35, 43A45, 47B47, 47D05, 54H20

ISBN 3-540-17199-1 Springer-Verlag Berlin Heidelberg New York
ISBN 0-387-17199-1 Springer-Verlag New York Berlin Heidelberg

Printing and binding: Druckhaus Beltz, Hemsbach/Bergstr.
2146/3140-543210

PREFACE

These lecture notes are based on a series of lectures given in a seminar at the Research Institute of Mathematical Sciences, Kyoto University, in 1984-85. I am greatly indebted to Huzihiro Araki for arranging the visit to Kyoto University, and thanks are due to the participants of the seminar for their interest.

The following colleagues gave valuable critical remarks to, and pointed out mistakes in, parts of previous versions of these notes: Huzihiro Araki, Charles J.K. Batty, George A. Elliott, David E. Evans, Akio Ikunishi, Akitaka Kishimoto and Derek W. Robinson. It is also a pleasure to thank Toshie Ito and the other "international" secretaries at RIMS, as well as Monica Grund in Trondheim, for their typing of various parts of the manuscript.

Trondheim, May 1986
Ola Bratteli

CONTENTS

Introduction with historical remarks

The theory of derivations on C*-algebras has developed in roughly three stages:

1. Bounded derivations: This subject was started in 1953 by Kaplansky, had its peak development around 1966-68 with the work by Kadison and Sakai, and was essentially finished in 1977-78 with work by Elliott, Akemann, and Pedersen, but the subject of bounded derivations from one algebra into another is still under development.

2. General theory of unbounded derivations: This subject was started by Sakai, Powers, Helemskii, Sinai, and Robinson around 1974, and had its maximal growth the following four years with contributions by Batty, Bratteli, Chi, Goodman, Herman, Jørgensen, Kishimoto, Lance, Longo, Mc Govern, Mc Intosh, Niknam, Ota, Takenouchi in addition to the five founders. This theory is concerned with the theory of closability, generator properties and classification of closed derivations, and was originally centered around the so-called Powers-Sakai conjecture. While the knowledge of the set of all closed derivations of a UHF algebra is still not much greater than it was soon after the founding of the subject, the classification problem of closed derivations on abelian C*-algebras is still developing, so far culiminating in Kurose's classification of the closed derivations of C([0,1]).

3. Non-commutative vectorfields: This subject came into being as a result of the merger of a question posed by Sakai at the Kingston conference in 1980 together with work pioneered by Batty, Goodman, and Nakazato on derivations on abelian C*-algebras invariant under an automorphism group, and the subject is also inspired by Connes work. At the present stage this theory concerns the classification and genrator properties of derivations which are well behaved with respect to the action of a locally compact group or Lie group in the sense that they commute or almost commute with the action and/or map classes of smooth elements with respect to the action into each other and/or has further properties of tangentiallity with respect to the action. In addition to the founders mentioned above this theory has been developed by Bratteli, Carey, Davies, Digernes, Elliott, Evans, Ikunushi, Jørgensen, Kishimoto, Kurose, Longo, Peligrad, Powers, Price, Robinson, Størmer, Takai, Takesaki, Thomsen, and Wassermann. This subject is still in vigorous development in several directions, and there are

both well-posed open problems within the present framework of the theory, and also the question of defining non-commutative manifolds in a more general fashion, without reference to a group action.

The main purpose of the present notes is to give a state of the art account of the third subject mentioned above, and trying to systematize the present knowledge. The knowledge of subjects 1 and 2 up to 1980 has already been treated fully in ([BR1], Chapter 3), ([Kad 1], articles by Sakai, Batty, Goodman, Takai and Bratteli-Jørgensen on p.p. 309-365) and in [Sak 5], and the more recent results on derivations on abelian C*-algebras has been treated in detail in lecture notes by Tomiyama, [Tom 1]. These results will therefore only be skimmed in the present notes, and no attempts to give complete references for the pre-1980 developments are done. On the other hand an attempt is made to give complete references for the subject 3, and the reference list contains some papers which are not used in the text, but which nevertheless are relevant for the subject.

Chapter 1 contains mostly standard material on the topics 1 and 2, and serves mainly to give outsiders a shorter introduction to the subject than in the available treatises. Except for a few date-ups in section 1.6, this material has been covered in more detail in the four references cited in the previous paragraph. Complete proofs are not always given in this chapter.

Chapter 2 is the main body of these lecture notes. Chapter 2.1 gives motivation and a rough survey of results, and may be read before Chapter 1.

Chapter 3 gives some results on the much less complete theory of dissipations.

More detailed historical remarks are given in the beginning of some of the proofs. Because of the unfinished state of this subject, several lemmas are stated and proved in more generality than actually needed for the present results, but these generalities will hopefully soon become useful.

The only prerequisite for reading these notes is a knowledge of elementary aspects of the theory of operator algebras and some harmonic analysis of abelian and compact groups. For later parts of the notes a knowledge of C^*-dynamical systems as exposed in [Ped 1] is useful, but all concepts used are defined here. On the other hand, these notes may be useful as an introduction to some concrete aspects of C^*-dynamical systems and "non-commutative differential geometry".

Chapter 1. General theory of derivations

§1.1 Basic Notions

Definition 1.1.1

Let A be a C*-algebra, and let δ be a linear operator from a dense *-subalgebra $D(\delta)$ of A, (called the domain of δ) into A.

(1) δ is called a *-derivation if

(1.1.1) $$\delta(x^*) = \delta(x)^* \qquad (x \in D(\delta)) \quad \text{and}$$

(1.1.2) $$\delta(xy) = \delta(x)y + x\delta(y) \qquad (x,y \in D(\delta)).$$

(2) δ is called a dissipation if

(1.1.3) $$\delta(x^*x) \leq \delta(x^*)x + x^*\delta(x) \qquad (x \in D(\delta)).$$

(Some authors use the converse inequality in this definition, and the term semiderivation is also used.)

(3) a dissipation δ is called a complete dissipation, if the matrix inequality

(1.1.4) $$[\delta(x_i^* x_j)] \leq [\delta(x_i^*)x_j + x_i^*\delta(x_j)]$$

is valid for all finite sequences $x_1 \cdots x_n \in D(\delta)$. Note that all derivations are complete dissipations, and all complete dissipations are dissipations.

Example 1.1.2

Let $t \in \mathbb{R} \mapsto \alpha_t$ be a strongly continuous one-parameter group of *-automorphisms of A. That is, α satisfies the following properties:

(1) for each $t \in \mathbb{R}$, α_t is a *-automorphism of A,

(2) $\alpha_{t+s} = \alpha_t \cdot \alpha_s \quad (t,s \in \mathbb{R})$,

(3) $\alpha_0 = \text{identity}$,

(4) $\|\alpha_t(x) - x\| \to 0 \ (t \to 0) \quad$ (for all $x \in A$).

Remark the fact that the conditáon (4) is equivalent to the norm continuity of $t \mapsto \alpha_t(x)$ for each x due to (2). The (infinitesimal) generator of α is defined by

(1.1.5) $D(\delta) = \{x \in A ; \lim_{t\to 0} \frac{1}{t} (\alpha_t(x)-x)$ exists in norm$\}$,

(1.1.6) $\delta(x) = \lim_{t\to 0} \frac{1}{t} (\alpha_t(x)-x) \qquad (x \in D(\delta))$.

Then δ is a *-derivation. The group $t \mapsto \alpha_t$ is norm continous, i.e. $\| \alpha_t - 1 \| \to 0$ $(t \to 0)$, if and only if δ is bounded, [BR1], Chapter 3.

Example 1.1.3

Let $L(A, A)$ be the set of all bounded linear operators on A. Let $t \in \mathbb{R}_+ = \{s \in \mathbb{R} ; s \geq 0\} \mapsto \alpha_t \in L(A, A)$ be a strongly continuous semi-group (i.e. $\alpha_{t+s} = \alpha_t \circ \alpha_s$ $(t,s \in \mathbb{R}_+)$, α_0 = identity and $t \to \alpha_t(x)$ is continuous for each $x \in A$), and assume that the generalized Schwarz's inequality

$$\alpha_t(x^*x) \geq \alpha_t(x)^* \alpha_t(x) \quad (x \in A, t \geq 0)$$

is satisfied for $t \geq 0$, $x \in A$.
If $-\delta$ is the generator of α (defined as in Example 1.1.2), and if x, and x^*x belongs to $D(\delta)$, then

$$\delta(x^*x) \leq \delta(x^*)x + x^*\delta(x).$$

In this case, $D(\delta)$ is not a *-subalgebra in general, [BR3].

§1.2 Bounded *-derivations

Definition 1.2.1

(1.2.1) $M(A) = \{y \in A^{**}; x \in A \Rightarrow xy, yx \in A\}$, where A^{**} is the Banach space bi-dual of A, is called the multiplier algebra of the C^*-algebra A.

By definition, if A has an identity, then $M(A)$ is nothing but A. A *-derivation δ is called inner of there exists an element h of

$M(A)$ such that

(1.2.2) $\delta(x) = [h, x] \quad (x \in D(\delta))$.

Remark 1.2.2

If A does not have an identity, then $M(A)$ is not A. For example, let $LC(H)$ be the C*-algebra of all compact linear operators on a Hilbert space H. In this case, $M(LC(H))$ is equal to $L(H)$, all bounded linear operators on H. For another example, let X be a locally compact Hausdorff space, and let $C_0(X)$ be the C*-algebra of all continuous functions on X vanishing at infinity. Then, $M(C_0(X))$ is equal to $C_b(X)$, all bounded continuous functions on X, [Ped 1].

In 1953, Kaplansky [Kap 1] showed that each *-derivation of a type I von Neumann algebra is inner. Sakai proved that an everywhere defined *-derivation is bounded in 1960 [Sak 1]. Sakai and Kadison obtained the result that each bounded *-derivation of a von Neumann algebra is inner in 1966 [Sak 2, Kad 2]. By this work, if δ is a bounded *-derivation of a C*-algebra A and (π, H) is a representation of A, then one can find an element h of $\pi(A)''$ such that $\delta(\cdot) = [h, \cdot]$. Sakai proved that one can choose $h \in M(A)$ if A is a simple C*-algebra in 1968, [Sak 3]. Around 1970, Elliott studied C*-algebras with outer (= not inner) bounded *-derivations, and this developed into the next theorem, which is due to Elliott (1977) [Ell 1], and Akemann-Pedersen (1978) [AP 1].

Theorem 1.2.3

Let A be a separable C*-algebra. Then the following conditions are equivalent.

(1) All bounded *-derivations of A are inner.

(2) Every summable central sequence in A is trivial.

(3) A is decomposable into a direct sum of two C*-algebras A_1 and A_2 : $A = A_1 \oplus A_2$, such that

(a) every central sequence in A_1 is trivial,

(b) A_2 is a (restricted) direct sum of simple C*-algebras.

The terms of this theorem are defined as follows:

Definition 1.2.4

Let $\{x_n\}_{n=1}^{+\infty}$ be a sequence of elements in a C*-algebra A.

(a) $\{x_n\}_{n=1}^{+\infty}$ is called central if $\{\|x_n\|\}_{n=1}^{+\infty}$ is uniformly bounded and $\|[x_n,y]\| \to 0$ $(n \to +\infty)$ for all $y \in A$.

(b) $\{x_n\}_{n=1}^{+\infty}$ is called summable if $\sum_n x_n$ converges in the strong topology of A^{**}.

(c) $\{x_n\}_{n=1}^{+\infty}$ is called trivial if there exists a sequence $\{z_n\}_{n=1}^{+\infty}$ in the center of $M(A)$ such that $\|(x_n - z_n)y\| \to 0$ $(n \to +\infty)$ for all $y \in A$.

Definition 1.2.5

Let $\{B_i\}_{i=1}^{+\infty}$ be a sequence of C*-algebras. Then their (restricted) direct sum is defined by

$$(1.2.3) \quad \bigoplus_{i=1}^{+\infty} B_i = \{\bigoplus_{i=1}^{+\infty} x_i \; ; \; x_i \in B_i, \; \|x_i\| \to 0 \quad (i \to +\infty)\}.$$

§1.3 Unbounded *-derivations

Sakai, Powers, Helemskii, Sinai and Robinson began investigating unbounded *-derivations around 1974 motivated by the study of time-development of quantum spin systems. These systems are described as follows.

Let ν be a positive integer. $L = \mathbb{Z}^\nu$ is called a ν-dimensional lattice. With each finite subset Λ in L, the C*-algebra

$$(1.3.1) \quad A_\Lambda = \bigotimes_{x \in \Lambda} A_x$$

is associated. Here A_x is the C*-algebra of all $n \times n$ matrices where n is a positive integer independent of the lattice site x. Then one has a natural identification

$$(1.3.2) \qquad \Lambda_1 \subset \Lambda_2 \Rightarrow A_{\Lambda_1} = A_{\Lambda_1} \otimes 1\!\!1_{\Lambda_2 \setminus \Lambda_1} \subseteq A_{\Lambda_2} .$$

A is defined to be C*-algebra inductive limit of A_Λ,

(1.3.3) $\mathcal{A} = \operatorname{ind}\lim_{\Lambda\to+\infty} \mathcal{A}_\Lambda \cong \overset{|\Lambda|}{\otimes} M_n$.

$\mathcal{A}$ is called a uniformly hyperfinite <u>(UHF)</u> <u>algebra</u> of <u>type</u> n^∞.

A function,

(1.3.4) Φ : X (a finite subset in L) $\mapsto \Phi(X) \in \mathcal{A}$

is called an <u>interaction</u> if $\Phi(X)$ is a self-adjoint element of $\mathcal{A}_X$ for each X. A formal Hamiltonian is defined by

(1.3.5) $H = \sum_{X\subset L} \Phi(X)$

and the sum is not well-defined in general, for example if Φ is nonzero and translationally invariant. Instead of considering the thermodynamic limit one can define the time-development of the infinite systems, by means of the *-derivation δ defined by Φ as follows:

(1.3.6) $\delta_\Lambda(x) = i[H_\Lambda, x] \qquad (x \in \bigcup_{\Lambda\subset L} \mathcal{A}_\Lambda)$,

where $H_\Lambda = \sum_{X\subset\Lambda} \Phi(X) \in \mathcal{A}_\Lambda$ If the condition,

(1.3.7) $\sum_{X\ni x} \| \Phi(X) \| < +\infty \qquad$ (for all $x \in L$)

is satisfied, then $\delta(x) = \lim_{\Lambda\to+\infty} \delta_\Lambda(x)$ exists and is a *-derivation defined on $\bigcup_\Lambda \mathcal{A}_\Lambda$.

<u>Example 1.3.1</u> (Ising Model in outer magnetic field)

Here n=2 and the interaction Φ is defined by

(a) $\Phi(X) = 0$ if $|X|$ =(the number of elements of X) ≥ 3 ,

(b) $\Phi(\{i\}) = h\sigma_z^{(i)}$ if $i \in L$,

(c) $\Phi(\{i,j\}) = \begin{cases} J\, \sigma_z^i \otimes \sigma_z^j & \text{if } i,j \in L \text{ and } |i-j| = 1, \\ 0 & \text{otherwise} . \end{cases}$

where $\sigma_z^{(i)} = \begin{pmatrix} 1 & 0 \\ 0 & -1 \end{pmatrix}$ on the lattice site i.

Example 1.3.2

An interaction Φ is called translationally invariant if

(1.3.8) $\Phi(X+x) = \tau_x\{\Phi(X)\}$ $(x \in L,\ X \subset L)$.

where τ_x is the automorphism of $\otimes M_n$ induced by translation by x on L.

An interaction Φ is said to have a finite range if there exists a $D \geq 0$ such that

(1.3.9) $\sup_{x,y \in X} |x-y| > D \Rightarrow \Phi(X) = 0$.

If Φ is a translationally invariant interaction and has a finite range, then δ is a pre-generator.
(A *-derivation δ is called pre-generator if δ is closable and its closure $\bar{\delta}$ is the generator of a one-parameter group of *-automorphism.) The derivation δ can be shown to be a pre-generator under much weaker conditions, see [BR 2], Theorem 6.2.4.

Next we discuss some general problems of (unbounded) derivations. The following problems are fundamental.

1. When is a derivation (pre) closed?
2. When is a derivation a (pre) generator?
3. Classify all (closed) derivations on a given C* algebra.

Now we discuss the first problem.

§1.4 Closed and pre-closed derivations

Definition 1.4.1

Let A be a Banach space and $D(\delta)$ be a subspace of A. A linear operator $\delta: D(\delta) \to A$ is closable (or preclosed) if it satisfies the following condition:

If $\{x_n\}_{n\in\mathbb{N}}$ is a sequence in $D(\delta)$, $\lim_{n\to\infty}\|x_n\| = 0$ and $\lim_{n\to\infty}\|\delta(x_n) - y\| = 0$ for some $y \in A$, then $y = 0$.

For a closable operator δ, the following extension $\bar{\delta}$, called the closure of δ, is well defined:

The domain $D(\bar{\delta})$ of $\bar{\delta}$ is the set of all $x \in A$ such that there exists a sequence $\{x_n\}$, $x_n \in D(\delta)$, and $y \in A$ satisfying $\lim x_n = x$, $\lim \delta(x_n) = y$, and then

(1.4.1) $\bar{\delta}(x) = y$.

We give some examples of closable and non closable operators.

Example 1.4.2

The generator of a semigroup of contractions is closed. See Theorem 1.5.1.

Example 1.4.3

An everywhere defined dissipation is closed. See Theorem 1.4.9.

Next is an example of a non-closable derivation.

Example 1.4.4

Let X be the Cantor set in [0, 1] and $C(X)$ be the set of all complex valued continuous functions on X. $C(X)$ is an abelian C*-algebra. Let δ be the first order differential operator defined in the obvious way on

(1.4.2) $D(\delta) \equiv \{f \in C(X);\ \delta(f)(x) \equiv \lim_{\substack{h\to 0\\ x+h\in X}} \frac{f(x+h) - f(x)}{h} \in C(X) \text{ exists}\}$.

δ is a non-zero derivation, because the function $f(x) = x$ is contained in $D(\delta)$ and $\delta(f) = 1$. We now show that $\delta = 0$ on a dense subset and hence δ is not closable.

Note that X has a basis consisting of open closed subsets. Then an arbitrary function $f \in C(X)$ may be approximated by a finite sum of projections in $C(X)$ as follows: Given $\varepsilon > 0$. For each point x, there exists an open closed neighbourhood N_x of x such that $|f(x') - f(x)| < \varepsilon$ for $x' \in N_x$. Since X is compact, a finite number of N_x, $x = x_1, \ldots, x_k$, will cover X and we may assume that the N_x are disjoint. Then $\operatorname{Sup}|f(x) - g_\varepsilon(x)| < \varepsilon$ for $g_\varepsilon \equiv \sum_i f(x_i)\chi_i$ where χ_i is the characteristic function of N_{x_i}, which belongs to $C(X)$. For any projection p in $C(X)$, which is nothing but the characteristic function of an open closed subset of X, $\delta(p)(x) = 0$. Therefore $\delta(g_\varepsilon) = 0$. Thus δ is not closable.

Example 1.4.4 may be extended to the case of a non-commutative C^*-algebra.

Example 1.4.5

Let A be the UHF algebra of Glimm type 2^∞: $A = \overset{\infty}{\otimes} M_2$, where M_2 is the algebra of all 2×2 matrices. We view A as the crossed product of the algebra $C(X)$ of Example 1.4.4 by a discrete group action as follows:

A point in X can be written uniquely as $\sum_{n=1}^{\infty} 3^{-n}c_n$ with $c_n = 0$ or 2 for each n and may thus be represented by a sequence $\{a_n\}_{n\ N}$ where $a_n = 0$ or 1 according as $c_n = 0$ or 2. So $X = \overset{\infty}{\times}\mathbb{Z}_2$. This identification is a homomorphism. Let G be the countable direct sum of the abelian group $\mathbb{Z}_2$, $G = \overset{\infty}{+}Z_2$. G has a natural action (by addition) on X and we denote the lifting of this action to $C(X)$ by α_g.

Let $\tilde{\delta}$ be the derivation of Example 1.4.4. Due to the fact $C(\mathbb{Z}_2)\times_\alpha Z_2 \cong M_2$, we have $A = C(X)\times_\alpha G$ where $C(X)\times_\alpha G$ denotes the C^* crossed product of $C(X)$ by the action (α, G). More precisely,

(1.4.3) $C(X) \times_\alpha G \equiv C^*$-closure of
$\{ \sum_{g\in G} x_g U_g;\quad x_g \in C(X),\ x_g = 0$ for all but a finite number of $g\}$.
where U_g $(g \in G)$ are unitaries satisfying the condition

(1.4.4) $U_g x U_g^* = \alpha_g(x) \quad (x \in C(X))$.

As $\alpha_g \tilde{\delta} = \tilde{\delta}\cdot\alpha_g$ on $C(X) \cap D(\tilde{\delta})$, the following defines a *derivation on A

(1.4.5) $\delta(\sum_g x_g U_g) = \sum_g \tilde{\delta}(x_g)\cdot U_g$,

Then

(1.4.6) $\delta|_{A_N} = 0$

where $A_N = \bigotimes_{n=0}^{N} M_2 \otimes 1$. Thus δ is zero on the dense subalgebra $\bigcup_N A_N$, and δ is not closable. This example is in [BR4].

Definition 1.4.6

A densely defined operator $-\delta$ on a Banach space A is called dissipative if one of the following equivalent conditions is satisfied:

(i) For each $x \in D(\delta)$, $x \neq 0$, there exists a nonzero $\eta \in A^*$ such that $\eta(x) = \|\eta\|\,\|x\|$ and $\mathrm{Re}\,\eta(\delta(x)) \geq 0$.
(A nonzero functional η satisfying the condition $\eta(x) = \|\eta\|\,\|x\|$ is called a tangent functional at x .)

(ii) For each $x \in D(\delta)$, $x \neq 0$, any tangent functional η at x satisfies the condtion $\mathrm{Re}\,\eta(\delta(x)) \geq 0$.

(iii) $\|(1 + \lambda\delta)(x)\| \geq \|x\|$ for any $x \in D(\delta)$ and for any $\lambda \in \mathbb{R}_+$.

We don't prove the equivalence of these conditions here. For the proof see [Bat 1], Lemma 1 and [DS 1], Theorem V.9.5.

Proposition 1.4.7

A dissipative operator is closable.

Proof

Let δ be a dissipative operator defined on $D(\delta)$. Let $\{x_n\}$ be a sequence in $D(\delta)$ such that $\lim_{n\to\infty} x_n = 0$ and $\lim_{n\to\infty} \delta(x_n) = y$.

If $x' \in D(\delta)$ and $\lambda > 0$, then by the condition (iii) of the above definition

$$(1.4.7) \quad \|(1+\lambda\delta)(x_n + \lambda x')\| \geq \|x_n + \lambda x'\|$$

By letting $n \to \infty$, we obtain

$$(1.4.8) \quad \|\lambda y + \lambda x' + \lambda^2 \delta(x')\| \geq \lambda\|x'\|$$

After dividing by λ, let $\lambda \to 0$. Then, $\|y + x'\| \geq \|x'\|$ for all $x' \in D(\delta)$. As $D(\delta)$ is dense, $y = 0$.

Example 1.4.8

Let S_t, $(t \in \mathbb{R}_+ \cup \{0\})$, be a strongly continuous semigroup of contractions on a Banach space and $-\delta$ its generator. Then $-\delta$ is dissipative.

Proof

Let $x \in D(\delta)$, $\eta(x) = \|\eta\|\cdot\|x\|$. Then

$$(1.4.9) \quad \eta(\delta(x)) = \lim_{t\to 0} \frac{1}{t}\eta(x - S_t(x)) = \lim_{t\to 0} \frac{\|\eta\|\ \|x\| - \eta(S_t(x))}{t}$$

As S is a contraction, $|\eta(S_t(x))| \leq \|\eta\|\cdot\|x\|$. So $\mathrm{Re}\ \eta(\delta(x)) \geq 0$.

Note that in this case

$$(1 + \lambda\delta)^{-1}(\cdot) = \int_0^\infty e^{-t} S_{\lambda t}(\cdot)dt$$

from which $\|(1 + \lambda\delta)^{-1}\| \leq 1$ is immediate.

Theorem 1.4.9 (Kishimoto)

Let A be a unital C^* algebra. Assume

(i) $D(\delta)$ is a dense* subalgebra of A ,

(ii) $\delta: D(\delta) \to A$ is a dissipation $(\delta(x^*x) \leq \delta(x^*)x + x^*\delta(x))$,

(iii) $\mathbb{1} \in D(\delta)$; if $x \geq 0$ and $x \in D(\delta)$ then $x^{\frac{1}{2}} \in D(\delta)$.

Then $-\delta$ is dissipative.

Remark

If δ is everywhere defined, the above condition (iii) is automatically satisfied.

Proof of Theorem 1.4.9

We follow the proof in [Kis 1].

Let $x \in D(\delta)$ and η be a tangent functional at x^*x satisfying $\|\eta\| = 1$. (Its existence is due to the Hahn Banach theorem.) Then η is a state: Consider the C*-subalgebra generated by x and 1. Then x is a function and η is a point measure at its maximum. Thus $\eta(1) = 1 = \|\eta\|$ which implies that η is a state of A .

Define $\eta_x (\in A^*)$ by $\eta_x(y) = \eta(x^*y)$. Then η_x is a tangent functional at x: We have $\eta_x(x) = \|x\|^2$ and $\|\eta_x\| \leq \|\eta\|\,\|x\| = \|x\|$, which imply $\|\eta_x\| = \|x\|$.

We now have

(1.4.10) $2\mathrm{Re}(\eta_x(\delta(x)) = \eta(x^*\delta(x)) + \eta(\delta(x^*)x) \geq \eta(\delta(x^*x))$

as η is a state. Define $y \equiv \sqrt{\|x^2\|\,\mathbb{1} - x^*x}$. Then $y \in D(\delta)$ by assumption (iii), and

(1.4.11) $2\mathrm{Re}(\eta_x(\delta(x))) \geq -\eta(\delta(y^2)) + \|x\|^2\eta(\delta(\mathbb{1}))$.

We have $\delta(\mathbb{1}) > 0$ because $\delta(\mathbb{1}) = \delta(\mathbb{1}^2) \leq 2\delta(\mathbb{1})$. Hence the second term is non-negative. On the other hand,

(1.4.12) $\eta(\delta(y^2)) \leq \eta(\delta(y)y) + \eta(y\delta(y))$

By $\eta(y^2) = 0$ and Schwarz inequality, the right hand side vanishes and $\eta(\delta(y^2)) \leq 0$. So $2\mathrm{Re}(\eta_x(\delta(x)) \geq 0$.

Corollary 1.4.10

An everywhere defined dissipation is bounded. (First adjoin 1 to A and extend the given dissipation δ by $\tilde{\delta}(1) = 0$ to a dissipation $\tilde{\delta}$ of $A + \mathbb{C}1 = \tilde{A}$. Then $\tilde{\delta}$ is dissipative by Proposition 1.4.9 and is closable by Proposition 1.4.7. Its closure is then bounded due to the closed graph theorem.)

Proposition 1.4.11

Let A be a C^* algebra and let $\delta: D(\delta) \to A$ be a *derivation. If there exists a family Ω of states of A such that $\omega\circ\delta \in A^*$ for all $\omega \in \Omega$ (this means that $|\omega\circ\delta(x)| \leq L_\omega \|x\|$ for some constant L_ω for all $x \in A$.) and $\bigoplus_{\omega\in\Omega} \pi_\omega$ is faithful where π_ω denotes the cyclic representation associated with ω, then δ is closable.

Proof

Let $\omega_{x,z}(y) \equiv \omega(x\ y\ z)$ for $x,z \in D(\delta)$. Then $\omega_{x,z}$ is in $D(\delta^*)$, the domain of the adjoint of δ, because

$$\omega(x\ \delta(y)z) = \omega(\delta(x\ y\ z)) - \omega(\delta(x)\ y\ z\ ) - \omega(x\ y\ \delta(z)) , \tag{1.4.13}$$

in which $(\omega\circ\delta)_{x,z}$, $\omega_{\delta(x),z}$ and $\omega_{x,\delta(z)}$ are all bounded functionals. Since

$$\omega_{x,z}(y) = (\pi_\omega(x^*)\Omega_\omega,\ \pi_\omega(y)\pi_\omega(z)\Omega_\omega) \tag{1.4.14}$$

and $\oplus\pi_\omega$ is faithful, $\{\omega_{x,z};\ x,z \in D(\delta)\}$ separates points in A and the linear hull of this set is weak*-dense. Therefore δ^* is weak*-densely defined and δ is closable. This proposition is due to Chi, [Chi 1].

1.5 Generators and pre-generators

Now we go into the second problem: When are derivations pre-generators? Recall the following definition

Definition 1.5.1

A densely defined operator $\delta: D(\delta) \to A$ on a Banach space A is said to be a generator if there exists a strongly continuous semigroup S of contractions on A (i.e. S is a map from $[0, \infty>$ into bounded operators on A such that $S_0 = 1$, $S_{t+s} = S_t S_s$, $\|S_t\| \leq 1$ and $t \to S_t x$ is norm continuous for each $x \in A$) such that δ is defined on $D(\delta) = \{x \in A \mid \lim_{t\to 0} \frac{1}{t}(S_t x - x)$ exists in norm$\}$ in the obvious manner. δ is a pregenerator if δ is closable and the closure $\bar{\delta}$ is a generator. The fundamental theorem on this subject is the following:

Theorem 1.5.2 (Hille- Yosida, Lumer-Phillips)

Let A be a Banach space, and $\delta: D(\delta) \to A$ a densely defined operator on A. The following conditions are equivalent.

(1) $-\delta$ is the generator of a strongly continuous semi-group of contractions on A.

(2) $-\delta$ is dissipative and

$$(1.5.1) \quad R(\mathbb{1} + \alpha\delta) \equiv (\mathbb{1} + \lambda\delta)D(\delta) = A .$$

for some (and then all) $\lambda > 0$.
(Note that if $-\delta$ is a generator, then $-\delta$ is closed by 2 and Definition 1.4.6, and a subspace $\mathcal{D} \subseteq D(\delta)$ is a core for δ if and only if $(1+\lambda\delta)\mathcal{D}$ is dense in A.)

Proof

(1 ⇒ 2) This follows from Example 1.4.8.
(2 ⇒ 1) One can use one of the following expressions to define the semigroup from

$$(1.5.2) \quad e^{-t\delta}x = \lim_{n\to\infty} (1+\tfrac{t}{n}\delta)^{-n}x = \lim_{\lambda\to 0} e^{-t\delta(1+\lambda\delta)^{-1}} x .$$

For example, the exponent of the last expression is bounded and hence the expressions is well-defined by power series. The first expression is well-defined since $\|(t+\frac{t}{n})^{-1}\| \leq 1$ by 2. For details see [BR 1, Theorem 3.1.10 and 3.1.16].

Definition 1.5.3

Let A be a Banach algebra. An element x of A is called analytic if $x \in D(\delta^n)$ $(n=1,2,\cdots)$, and

$$(1.5.3) \qquad \sum_{n=0}^{+\infty} \frac{t^n}{n!} \|\delta^n(x)\| < +\infty \quad \text{for some positive } t.$$

Theorem 1.5.4 (Nelson)

Let A be a Banach space, and $\delta : D(\delta) \to A$ a densely defined closed operator on A. The following conditions are equivalent

(1) δ generates a strongly continuous one-parameter group of isometries.

(2) $\pm\delta$ are dissipative and δ has a dense set of analytic elements.

Proof :

$(1 \Rightarrow 2)$ By the preceding theorem, $\pm\delta$ is dissipative. Put $\alpha_t = \exp(t\delta)$. For an arbitary element $x \in A$, define

$$(1.5.4) \qquad x_n = \sqrt{\frac{n}{\pi}} \int_{-\infty}^{+\infty} e^{-nt^2} \alpha_t(x)\, dt .$$

Then, x_n is entire analytic, and the sequence $\{x_n\}_{n=1}^{\infty}$ converges to x. For the analyticity, note that

$$(1.5.5) \qquad \delta(\alpha_f(x)) = -\alpha_{f'}(x)$$

where

$$(1.5.6) \qquad \alpha_f(x) = \int_{-\infty}^{+\infty} f(t)\alpha_t(x)dt.$$

$(2 \Rightarrow 1)$ Exponentiate and use dissipativeness to controll norms. For details, see [BR 1], Theorem 3.1.19.

Example 1.5.5

The existence of a dense set of analytic elements alone, without assuming dissipativity of $\pm\delta$ does not insure that δ is a generator. For example, let A be the C*-algebra $C([0,1])$, and $\delta = \frac{d}{dx}$. Then each polynomial is entire analytic for δ, but $\sum_{n=0}^{+\infty} \frac{t^n}{n!}(\frac{d}{dx})^n(x^m) = (x+t)^m$, therefore $e^{t\delta}$ is not bounded, and δ is not a generator. It is easily verified directly that δ and $-\delta$ are not dissipative.

Corollary 1.5.6

Let A be a C*-algebra, and $\delta : D(\delta) \to A$ a *-derivation. Assume that there exists a family Λ of states on A such that

$$(1.5.7)\begin{cases} \omega\circ\delta = 0\,, & (\omega \in \Lambda), \\ \bigoplus_{\omega\in\Lambda} \pi_\omega & \text{is faithful} \end{cases}$$

where π_ω is the cyclic representation of A associated to ω ([BR 1], Definition 2.3.18). Then δ is closable.

Moreover assume either that $R(\mathbb{1} + \alpha\delta)$ is dense in A for all $\lambda \in \mathbb{R}$, or that δ has a dense set of analytic elements. Then δ is a pre-generator of a strongly continuous group of *-automorphisms.

Proof

The derivation δ is closable by Proposition 1.4.11, and replacing δ by its closure we have to show only that $\pm\delta$ are dissipative. For simplicity, assume that $\Lambda = \{\omega\}$ consists of one element. Let (π, H, Ω) be the cyclic representation associated to ω, and define tentatively an operator H on the set $\pi(D(\delta))\Omega$ by

$$(1.5.8.)\quad H\pi(x)\Omega = \pi(\delta(x))\Omega \qquad (x \in A).$$

Then, for $x, y \in D(\delta)$ $\quad$ ($D(\delta)$ is dense in A),

$$(1.5.9.)\quad 0 = \omega(\delta(y^*x)) = \omega(\delta(y^*)x) + \omega(y^*\delta(x))$$

$$= (H\pi(y)\Omega, \pi(x)\Omega) + (\pi(y)\Omega, H\pi(x)\Omega).$$

By this equation, (1) H is well defined because $\pi(x)\Omega = 0$ implies $H\pi(x)\Omega = 0$, and, (2) H is skew symmetric ($H^* \supset -H$) and hence essentially skew adjoint by each of the two alternative assumptions. By the relation,

$$(1.5.10)\quad \pi(\delta(x))\pi(y)\Omega = \pi(\delta(x)y)\Omega = \pi(\delta(xy) - x\delta(y))\Omega$$

$$= (H\pi(x) - \pi(x)H)\pi(y)\Omega \qquad (x,\ y \in D(\delta))\ ,$$

we obtain $\pi(\delta(x)) = H\pi(x) - \pi(x)H$. Let $V_t = e^{t\bar{H}}$ be the unitary group generated by the skew-adjoint closure $\bar{H}$ of H.
Then $\pi(\delta(x))$ is the derivative of $V_t\pi(x)V_t^*$ at $t=0$. This implies that

$$(1.5.11)\qquad \|(\mathbb{I} + \lambda\delta)(x)\| = \|\pi(1 + \lambda\delta)x)\| \geq \|\pi(x)\| = \|x\|$$

for $x \in D(\delta)$, $\lambda \in \mathbb{R}$. Therefore $\pm\delta$ are dissipative and δ is a generator of a one-parameter group of isometries. Using analytic elements and the derivation property of δ, these isometries are *-automorphisms.

Example 1.5.7

Let $A = \bigotimes_{x \in L} M_n$ be the C*-algebra of a quantum lattice system. (See Examples 1.3.1 and 1.3.2). Let ϕ be an interaction satisfying,

$$(1.5.12)\qquad \|\Phi\|_\lambda \equiv \sum_{n=0}^{+\infty} e^{\lambda n} \Big(\sup_{x \in L} \sum_{\substack{X \ni x \\ |X| = n+1}} \|\Phi(X)\|\Big) \quad < +\infty$$

for some $\lambda > 0$. Let ω be the unique tracial state on A. Then, as $\omega(\delta_\Lambda(x)) = 0$ for all finite subsets Λ of L and $x \in \bigcup_\Lambda A_\Lambda$, we have

$$(1.5.13)\qquad \omega(\delta(x)) = 0 \qquad (x \in \bigcup_\Lambda A_\Lambda).$$

Hence δ is closable. Moreover, an explicit calculation shows that $\bigcup_\Lambda A_\Lambda$ consists of analytic elements for $\bar{\delta}$. The above Corollary then implies that δ is pre-generator. (See [BR 2], Theorem 6.2.4. for details)
In this case, $e^{t\delta}(x) = \lim_{\Lambda\to\infty} e^{t\delta_\Lambda}(x) \quad (x \in A)$, uniformly for t in compacts.
This follows from the next theorem.

Theorem 1.5.8 (Kurtz)

Let $\{-\delta_n\}_{n=0}^{+\infty}$ be a sequence of generators of contraction semi-groups on a Banach space A. Let $G(\mathbb{1} + \lambda\delta_n)$ be the graph of $(1 + \lambda\delta_n)$. Let $G_\lambda \equiv \lim_{n\to\infty} G(1+\lambda\delta_n)$ be the limit graph. The following are equivalent.

(1) There exists a strongly continuous semi-group S of contractions such that

$$(1.5.14) \qquad \lim_{n\to+\infty} \|e^{-t\delta_n}(x) - S_t(x)\| = 0 \quad (\text{for all } x \in A, \ t \in \mathbb{R}_+)$$

(where the convergence is uniform for t in compacts).

(2) The domain $D(G_\lambda)$ and the range $R(G_\lambda)$ are norm-dense in A for some $\lambda > 0$.

In this case, $G_\lambda = G(\mathbb{1} + \lambda\delta)$, where $-\delta$ is the generator of S_t.

Proof.

See [BR 1], Theorem 3.1.28.

The following corollary is often useful in applications, and implies the last statement in Example 1.5.7.

Corollary 1.5.9

Let $-\delta_n$ be the generator of semi-groups of contractions $e^{-t\delta_n}$ on A. Assume that there exists a dense subspace $D(\delta)$ in A such that

$$(1.5.15) \qquad \delta(x) \equiv \lim_{n\to\infty} \delta_n(x)$$

exists for all $x \in D(\delta)$. If the range $R(1 + \alpha\delta)$ is dense in A for some $\alpha > 0$, then $-\bar{\delta}$ is the generator of a semi-group of contractions and

$$(1.5.16) \qquad e^{-t\bar{\delta}}(x) = \lim_{n\to\infty} e^{-t\delta_n}(x).$$

for $x \in A$, $t \geq 0$, where the convergence is uniform on bounded subsets of t in $\mathbb{R}_+$.

§1.6 Classification of all closed derivations on a given C*-algebra.

The most important technique in this connection is :

§1.6.1 Functional analysis of the domains of derivations.

We start with an example :

Example 1.6.1

Let A be an abelian C*-algebra and $\delta : D(\delta) \to A$ a closed *derivation. If $f \in D(\delta)$ and $g(t) \in C^1(\mathbb{R})$, then $g(f) \in D(\delta)$ and

$$\delta(g(f)) = g'(f)\,\delta(f) \tag{1.6.1}$$

where $g'(t)$ is the derivative of the function $g(t)$, and $C^1(\mathbb{R})$ is the space of once continuously differentiable functions on $\mathbb{R}$. Actually (1.6.1) can easily be checked for $g(t) = t^n$ and hence for $g(t) = P(t)$, a polynomial. By Stone-Weierstrass's theorem, there exists a sequence of polynomials $\{P_n(t)\}_{n \in N}$ such that

$$\sup_{|t| \leq \|f\|} |g(t) - P_n(t)| \to 0 \tag{1.6.2}$$

$$\sup_{|t| \leq \|f\|} |g'(t) - P'_n(t)| \to 0 \tag{1.6.3}$$

as $n \to \infty$.
Then $g(f) = \lim_{n\to\infty} P_n(f)$ and $\lim_{n\to\infty} \delta(P_n(f)) = \lim_{n\to\infty} P'_n(f)\delta(f) = g'(f)\delta(f)$. By closedness of δ it follows that $g(f) \in D(\delta)$ and $\delta(g(f)) = g'(f)\delta(f)$.
The formula (1.6.1) may be rephrased as follows: $D(\delta)$ is closed under C^1functional analysis. This is not longer true when A is not abelian. For the general case we have the following.

Theorem 1.6.2 (Powers)

Let δ be a closed *-derivation. If f is a real function satisfying

(1.6.5) $\int_{-\infty}^{\infty} |p|\,|\hat{f}(p)|\,dp < \infty$

where

(1.6.6) $\hat{f}(p) = \frac{1}{\sqrt{2\pi}} \int e^{-itp} f(t)dt$,

and x is a selfadjoint element in $D(\delta)$, then $f(x) \in D(\delta)$ and

(1.6.7) $\delta(f(x)) = \frac{1}{\sqrt{2\pi}} \int_{-\infty}^{\infty} ip\hat{f}(p)\ (\int_0^1 e^{itpx}\ (x)e^{i(1-t)px}\,dt)dp$

where

(1.6.8) $f(x) = \frac{1}{\sqrt{2\pi}} \int_{-\infty}^{\infty} \hat{f}(p)e^{ipx}\,dp$.

The proof is carried out in three steps.

Step 1. Let $x = x^* \in D(\delta)$ and $\lambda \notin \mathrm{Spec}(x)$. Then $x(\lambda 1 - x)^{-1} \in D(\delta)$ and

(1.6.9) $\delta(x(\lambda\mathbb{1} - x)^{-1}) = \lambda(\lambda\mathbb{1} - x)^{-1}\delta(x)(\lambda\mathbb{1} - x)^{-1}$

The proof is by series expansion. See [BR 1], Proposition 3.2.29. Note that $x(\lambda\mathbb{1} - x)^{-1} \in A$ even if $\mathbb{1} \notin A$.

Step 2. If $\mathbb{1} \in A$, then $\mathbb{1} \in D(\delta)$.

Proof

As $D(\delta)$ is dense, it contains a positive invertible element X . (Take an element in $D(\delta)$ close to 1 and average with its adjoint.) Then

(1.6.10) $\delta(x\frac{\mathbb{1}}{\varepsilon\mathbb{1}+x}) = -\varepsilon\frac{\mathbb{1}}{\varepsilon\mathbb{1}+x}\,\delta(x)\frac{\mathbb{1}}{\varepsilon\mathbb{1}+x}$

by Step 1, and thus

(1.6.11) $\|\delta(x\frac{\mathbb{1}}{\mathbb{1}+x})\| \leq \varepsilon\,\|\frac{\mathbb{1}}{x}\|^2\,\|\delta(x)\|$.

Thus $\delta(x(\varepsilon\mathbb{1} + x)^{-1}) \to 0$ as $\varepsilon \to 0$.

By the closedness of δ and by $x\frac{\mathbb{1}}{\varepsilon+x} \to \mathbb{1}$ as $\varepsilon \to 0$, we obtain $\mathbb{1} \in D(\delta)$ and $\delta(\mathbb{1}) = 0$.

Step 3. If $x=x^* \in D(\delta)$ then $e^{ipx} \in D(\delta)$ for $p \in \mathbb{C}$ and

(1.6.12) $$\delta(e^{ipx}) = ip \int_0^1 e^{itpx}\, \delta(x)\, e^{i(1-t)px}\, dt.$$

Proof We use

(1.6.13) $$e^{ipx} = \lim_{n\to\infty} (\mathbb{1} - \frac{ipx}{n})^{-n} .$$

As $(\mathbb{1} - \frac{ipx}{n})^{-1} = \mathbb{1} + \frac{ipx}{n}(\mathbb{1} - \frac{ipx}{n})^{-1}$, Steps 1 and 2 imply $(\mathbb{1}-\frac{ipx}{n})^{-1} \in D(\delta)$ and

(1.6.14) $$\delta((\mathbb{1} - \frac{ipx}{n})^{-n}) = \sum_{m=0}^{n} (\mathbb{1} - \frac{ipx}{n})^{-m} \left[(1 - \frac{ipx}{n})^{-1} \frac{ip}{n}\delta(x)(1-\frac{ipx}{n})^{-1}\right] \times (1 - \frac{ipx}{n})^{-(n-m-1)}$$

$$= ip \sum_{m=0}^{n} \frac{1}{n}\left[(1 - \frac{ipx}{n})^{-n}\right]^{(\frac{m+1}{n})} \delta(x) \left[(1 - \frac{ipx}{n})^{-n}\right]^{(1-\frac{m+1}{n})}$$

$$\xrightarrow[n\to\infty]{} (ip)\int_0^1 e^{itpx}\delta(x)\, e^{i(1-t)px}\, dt .$$

By closedness of δ we have the Theorem.

Remark 1.6.3

The condition $\int |p||\hat{f}(p)|dp < \infty$ is fullfilled for $f \in C^2(\mathbb{R})$ or for $f \in C^{1+\varepsilon}(\mathbb{R})$ but not for $f \in C^1(\mathbb{R})$. A counterexample (due to McIntosh [McI]) is given by

(1.6.15) $$f(t) = |t|\, \{ \log|\log|t|| \}^{-\alpha} \quad (0< \alpha <1) .$$

This function is continuously differentiable at $t=0$ but there is an explicit example of a closed derivation δ and an $x=x^* \in D(\delta)$ such that $f(x) \notin D(\delta)$ for this f.

§1.6.2. Classification of the closed derivations on the compact operators

Example 1.6.4 (Classification of all closed derivations of $LC(H)$ $\equiv$ the set of all compact operators on H.)

The following conditions are equivalent.

1. δ is a closable derivation.
2. There exists a skew symmetric operator H, i.e. $H \subseteq -H^*$, such that $\delta \subseteq \mathrm{ad}(H)$.

Proof

2 ⇒ 1. This follows from the closedness of H.

1 ⇒ 2. Assume that δ is closed. Let p be a one dimensional projection. Then for any $\varepsilon > 0$ there exists a $x = x^* \in D(\delta)$ such that $\|x - p\| \leq \varepsilon$. But as p is a projection, we have $\|x^2 - x\| \leq 3\varepsilon + \varepsilon^2$, and hence $\mathrm{Spec}\,(x) \subseteq [-\varepsilon', \varepsilon'] \cup [1-\varepsilon', 1+\varepsilon']$ where $\varepsilon' = (1 - \sqrt{1-(12\varepsilon+4\varepsilon^2)})/2 \approx 3\varepsilon$. Choose ε sufficiently small that $2\varepsilon' < 1$, and let $f(t)$ be a smooth function satisfying

$$(1.6.16)\quad f(t) = 0 \ \text{ if } \ |t| < \varepsilon', \quad f(t) = 1 \ \text{ if } \ |t - 1| < \varepsilon'.$$

Then $e = f(x)$ is a projection and $e \in D(\delta)$ by Theorem 1.6.2. Furthermore

$$(1.6.17)\quad \|e - p\| \leq \|e - x\| + \|x - p\| \leq 2\varepsilon' < 1.$$

Hence e is 1 dimensional. Set

$$(1.6.18)\quad \delta_e \equiv \mathrm{Ad}(e\delta(e) - \delta(e)e), \quad \delta^e \equiv \delta + \delta_e.$$

We have

$$(1.6.19)\quad \delta^e(e) = \delta(e) + e\delta(e)e - \delta(e)e - e\delta(e) + e\delta(e)e$$

$$= \delta(e-e^2) + 2e\delta(e)e = 2e\delta(e)e = 0$$

because

$$(1.6.20)\quad e\delta(e)e = e(\delta(e)e + e\delta(e) - \delta(e))e = e\delta(e^2-e)e = 0.$$

As δ^e is an inner perturbation of δ, we may from now on assume $\delta(e) = 0$ for our purpose.

Let ψ be a unit vector in the range of e,

(1.6.21) $eH = \mathbb{C}\psi$.

We define an operator H by

(1.6.22) $Hx\psi \equiv \delta(x)\psi \quad (x \in D(\delta))$.

It is proved as in the argument for Corollary 1.5.6 that H is well defined and

(1.6.23) $H \subseteq -H^*$, $\delta(x) = \mathrm{ad}(H)(x) \quad (x \in D(\delta))$.

§1.6.3 Classification of the closed derivations on an AF algebra

Definition 1.6.5

A C*-algebra A is approximately finite dimensional, AF, if there exists a sequence of finite dimensional *subalgebras $\{A_n\}$ such that $A_n \subset A_{n+1}$ for all n and $\bigcup_n A_n$ is dense in A, see [Bra 1] .

Theorem 1.6.6 (Sakai)

Let A be an AF algebra, δ be a closed derivation on A . Then there exists an increasing sequence of finite dimensional *subalgebras $\{A_n\}$ such that $\bigcup_n A_n \subseteq D(\delta)$ and $\bigcup_n A_n$ is dense in A . If δ is a generator, $\{A_n\}$ can be chosen such that $\bigcup_n A_n$ consists of analytic elements for δ .

Proof

Combine Glimm's techniques for UHF-algebras with functional analysis of the domain, see [Sak 4] .

Remark 1.6.7 (Elliott)

This is a partial classification result because there exists $h_n = -h_n^* \in A$ such that

(1.6.24) $\quad \delta|_{A_n} = \mathrm{ad}(h_n)|_{A_n}$.

In fact, let $\{e^m_{ij}\}$ be a set of matrix units for the finite-dimensional algebra A_n , i.e.

$$e^{(m)}_{ij}\, e^{(s)}_{k\ell} = \delta_{m,s}\, \delta_{j,k}\, e^{(m)}_{i\ell}\ , \quad e^{(m)*}_{ij} = e^{(m)}_{ji}\ .$$

Then $h_n \equiv \sum_{m,i} \delta(e^{(m)}_{i1})\, e^{(m)}_{1i}$ has the desired property.

(h_n is skew adjoint due to $\sum_{i,m} e^{(m)}_{i1}\, e^{(m)}_{1i} = \mathbf{1}$.)

<u>Problem 1.6.8</u> (Powers - Sakai conjecture)

Can $\bigcup_n A_n$ be taken to be a core for δ when δ is a generator? By Remark 1.6.7. and Corollary 1.5.9 this would imply that all one-parameter groups $t \to \alpha_t$ of *-automorphisms of an AF algebra is approximately inner in the sense that there would exist a sequence $h_n = -h_n^* \in A$ such

$$\lim_{n\to\infty} \| \alpha_t(x) - e^{th_n} x e^{-th_n} \| = 0$$

for all $x \in A$, $t \in \mathbb{R}$, uniformly for t in compacts. It is easily shown that a single automorphism α of an AF-algebra is approximately inner provided it preserves equivalence classes of projections. The solution of this problem has consequences for the theory of equilibrium states of quantum lattice systems, see [BR 2], Example 5.3.26.

§1.6.4 <u>Classification of closed derivations on abelian C*-algebras.</u>

By Gelfand Theory, a general abelian C*-algebra has the form $A = C_0(X)$, where X = spectrum of A, and $C_0(X)$ is the algebra of continuous functions on X vanishing at infinity.

By the dimension of a locally compact space X, we mean the Hausdorff dimension of X. The known classification results for derivations depends heavily on the dimension on X.

In the case that $\dim X = 0$ it turns out that all closed derivations are trivial if X is total disconnected, as we will see. For the case that $\dim X = 1$ a complete classification for $A = C([0,1])$ was done by Kurose. For more than 2 dimensions only sporadic results are known.

§1.6.4.1 The 0-dimensional case.

Lemma 1.6.9

Let A be an abelian C*-algebra, δ a closed derivation, and e be a selfadjoint projection in A.
Then $e \in D(\delta)$ and $\delta(e) = 0$.

Proof

By functional analysis of the domain as in Example 1.6.4., there exists a projection p such that $p \in D(\delta)$ and $\| p - e \| < 1$. But as A is abelian it follows that $p = e$. By (1.6.20) $e\delta(e) = e\ \delta(e)e = 0$. Replacing e by $(1-e)$ and using $\delta(1)=0$, we obtain $(1-e)\delta(e) = 0$. Hence $\delta(e) = 0$.

We will consider abelian AF algebras, which are characterized as follows (strictly speaking we do not need the following proposition; it is stated to make the connection with 0-dimensional spaces).

Proposition 1.6.10

Let A be an abelian C*-algebra, and X be the spectrum of A. (i.e. X is the locally compact Hausdorff space of maximal ideals in A with the Jacobson topology, and then $A = C_0(X)$.) The following conditions are equivalent.

(1) A is an AF-algebra.

(2) A is generated by a sequence of projections in the norm topology.

(3) X is second countable and totally disconnected (i.e. the connected component of each point of X consists of the point itself).

(4) X is homeomorphic to a totally disconnected closed subset of $\mathbb{R}$.

(5) X is a second countable space with a basis of open compact sets.

Proof

See [Bra 2], Proposition 3.1, and [BEℓ 1].

Corollary 1.6.11

An abelian AF-algebra admits no nonzero closed derivations.

Proof

By Lemma 1.6.9, if e A is a self-adjoint projection, then e $D(\delta)$ and $\delta(e) = 0$. Hence by Proposition 1.6.10, δ is zero on a dense subset of A.

This result can be extended slightly.

Theorem 1.6.12 (Batty)

If X is a compact Hausdorff space with a dense, open, totally disconnected subset, then $C(X)$ admits no nonzero closed derivations.

Proof

See [Bat 2].

Remark 1.6.13 (Batty)

Any second countable compact Hausdorff space X can be embedded in a space with a dense, open, totally disconnected, countable subspace Y, by adding Y as a sequence of points such that each point of X is a limit point of the sequence Y. Thus the class of spaces covered by Theorem 1.6.12 may have arbitrarily high dimension. Thus the heading "The 0-dimensional case" of this subsection only applies to perfect spaces, where any two neighbourhoods of any two points has homeomorphic sub-neighbourhoods.

There exists a connected, locally pathwise connected, compact Hausdorff space X such that $C(X)$ admits no nonzero closed derivations. (In this example, X is non-metrizable, and for each $f \in C(X)$, $X_f = \{x \in X;\ f$ is constant in a neighbourhood of $x\}$ is dense in X.) See [Bat 3].

Problem 1.6.14

When X is a second countable compact Hausdorff space, are the following equivalent?

(1) C(X) admits no nonzero closed derivations.

(2) X contains a dense, open, totally disconnected subset.

(Separability of X is necessary, see Remark 1.6.13.)

§1.6.4.2 The 1-dimensional case

We will consider the unit interval $X = I \equiv [0, 1]$.

Definition 1.6.15

Let A be a C*-algebra, and δ be a densely defined derivation. A self-adjoint element x of A is called well-behaved if there exists a state ω such that

(1.6.25) $|\omega(x)| = \|x\|$,

(1.6.26) $\omega(\delta(x)) = 0$.

The derivation δ is called conservative if all self-adjoint elements in $D(\delta)$ are well-behaved. (Note that δ is well-behaved if and only if $\pm\delta$ is dissipative, see Proposition 2.3.4.) δ is called quasi conservative if the interior of the set of all well-behaved elements is dense in $D(\delta)$. The first idea in classifying derivations δ on C[0, 1] is of course to relate δ to $\frac{d}{dx}$. A result in this direction is

Theorem 1.6.16 (Batty-Goodman-Sakai)

Let δ be a closed derivation on the C*-algebra $C(I)$. Then the following are equivalent.

(1) δ is quasi conservative.

(2) $D(\delta)$ contains a strictly monotone function.

(3) There exists a function $\lambda \in C(I)$ and a homeomorphism θ of $[0, 1]$ (i.e. an automorphism of $C(I)$) such that,

$$(1.6.27) \quad \delta \supseteq \theta\lambda\frac{d}{dx}\,\theta^{-1} ,$$

where the domain of the latter operator is defined in the obvious manner so that it is closed.

Proof

See [Bat 4], [Goo 1], [Sak 5].

Remark 1.6.17

There are examples where

$$(1.6.28) \quad \delta \not\supseteq \theta\lambda\frac{d}{dx}\,\theta^{-1} ,$$

see e.g. Theorem 1.6.21.

Because of Remark 1.6.17, the classification in Theorem 1.6.16 is only partial. However, a complete classification has been obtained by Kurose, and the first step is the following theorem.

Theorem 1.6.18 (Kurose)

The following conditions are equivalent.

(1) δ is a closed derivations of $C(I)$ with the range $R(\delta) = C(I)$ and the kernel $\operatorname{Ker}\delta = \mathbb{C}\cdot \mathbb{1}$.

(2) There exists a non-atomic signed measure μ on I with $\operatorname{supp}\mu = [0, 1]$ such that

$$(1.6.29) \quad D(\delta) = \{g \in C(I);\ \exists f \in C(I),\ \exists \lambda \in \mathbb{C}\ \text{ s.t. }\ g(x) = \lambda + \int_0^x f\,d\mu\},$$

$$(1.6.30) \quad \delta(g) = f .$$

Proof

See [Kur 1], Theorem 2.3 or [Tom 1], Theorem 3.1.3.

Remark 1.6.19

When the derivation δ is given, the measure μ satisfying the above condition is uniquely determined. There exists a dense subset $U \subsetneq I$ such that

$$(1.6.31)\quad \delta(g)(x) = \lim_{h\to 0} \frac{g(x+h) - g(x)}{\mu(x+h) - \mu(x)} \quad (x \in U,\ g \in D(\delta)),$$

where $\mu(x) = \int_0^x d\mu$, so we put $\delta = \frac{d}{d\mu}$.

More generally, let (A, B, μ) be a triplet such that

$$(1.6.32)\quad \begin{cases} A \text{ is an open subset of } I, \\ B \text{ is a dense open subset of } A, \\ \mu \text{ is a continuous function on } B \text{ which has a finite} \\ \text{total variation on each closed subinterval } E \subsetneq B. \end{cases}$$

For each such E, define the derivation $\delta_{\mu|E}$ on $C(E)$ as in the preceding theorem. If f is an element of $C(I)$ and $f_{|E}$ belongs to $D(\delta_{\mu|E})$ for all such E, then define

$$(1.6.33)\quad \frac{df}{d\mu}(x) = \begin{cases} \delta_{\mu|E}(f)(x) & (\text{if } x \in E \subsetneq B), \\ 0 & (\text{if } x \notin A), \end{cases}$$

This definition can be shown to be independent of the particular E which is chosen for a given x. Define

$$(1.6.34)\quad D(\delta(A, B, \mu)) = \{f \in C(I);\ \frac{df}{d\mu} \text{ exists and has a continuous extension to } I\},$$

$$(1.6.35)\quad \delta(A, B, \mu)(f) = \text{this continuous extension of } \frac{df}{d\mu}.$$

Theorem 1.6.20 (Kurose)

(1) $\delta(A, B, \mu)$ is a closed derivation.

(2) Conversely, if δ is a closed derivation on $C(I)$, then there exists a triple (A, B, μ) satisfying (1.6.32) and

(1.6.36) $\delta \subsetneq \delta(A, B, \mu)$.

Proof

See [Kur 2], Theorem 6 and [Kur 3], Theorem 1, or [Tom 1], Theorem 3.2.2 and Theorem 3.2.3. Although Theorem 1.6.20 gives a complete classification of closed derivations of $C(I)$, some of the structure of these derivations is not obvious from the theorem, as the following surprising result shows:

Theorem 1.6.21 (Tomiyama)

Each closed nonzero derivation on $C(I)$ has a proper closed extension.

Proof

See [Tom 2], Theorem 1.3.

§1.6.4.3 The 2-dimensional case

In this case, several attempts has been made to prove that a closed derivation at a generic point is given by a differentiation in some direction in the sense of Remark 1.6.19, but so far without success. We will mention a partial result in this vein, however, for this, we need a concept which is useful also in other connections.

Definition 1.6.22 (Batty)

Let δ be a derivation on an abelian C*-algebra $C_0(X)$. A closed subset $E \subseteq X$ is called self-determining for δ or a restriction set if $f \in D(\delta)$ and $f|_E = 0$ imply $\delta(f)|_E = 0$. See [Bat 4], [Goo 1].

Example 1.6.23

Let δ be a generator of a flow T_t on X;

(1.6.37) $e^{t\delta}(f)(x) = f(T_t x) \qquad (x \in X, \quad f \in C(X))$.

(A flow is a continuous one-parameter group of homeomorphisms, which via Gelfand theory is the same as a strongly continuous group of *-automorphisms, see Lemma 2.4.1 and Definition 2.4.2)

For a compact subinterval E of $\mathbb{R}$ and $x \in X$, the orbit segment $T_E x = \{T_t x ; t \in E\}$, which is tangential to the flow, is self-determining. On the other hand, one easily sees that orbit segments transversally to the flow are not self-determining.

Observation 1.6.24

When $E \subseteq X$ is self-determining, the relation

$$(1.6.38) \quad \delta_E(f|_E) = \delta(f)|_E$$

defines a derivation δ_E on $\{f|_E ; \delta \in D(\delta)\}$.

Lemma 1.6.25 (Batty)

Let δ be a closed derivation, and U a open set of X. Then $\overline{U}$ is self-determining and $\delta_{\overline{U}}$ is closable.

Proof

Use functional analysis of the domain, see [Bat 4], Lemma 4.1. For definiteness, we now consider $x = I \times I = [0, 1] \times [0, 1]$. The partial result alluded to is:

Theorem 1.6.26 (Nishio)

Let δ be a closed derivation on $C(I \times I)$. Assume that the range of δ coincides with $C(I \times I)$. Then for any open subset $V \subseteq I \times I$, there exists a non-empty, open and connected subset $U \subseteq V$ such that $\operatorname{Ker} \delta_{\overline{U}}$ contains a non-constant function in $D(\delta_{\overline{U}})$.

Proof

See [Nis 1].

Remark 1.6.27

Note that we cannot expect δ to have a global kernel property analogous to Theorem 1.6.26, because δ could be the generator of an ergodic flow on $I \times I$, i.e. $\operatorname{Ker}(\delta) = \mathbb{C}\mathbb{1}$.

Chapter 2. Noncommutative vectorfields

§2.1 General introduction and motivation

We have seen in the previous sections that a complete classification of all closed derivations of a C*-algebra A only has been made in some ultraspecial cases:

$A = LC(H)$

A is an abelian AF-algebra, or the spectrum of A has very special properties.

$A = C([0,1])$

(Presumably one can combine the two first cases and classify all closed derivations of type I AF algebras with continous trace, more interesting would be to combine the first and third case and classify all closed derivations of $LC(H) \otimes C([0,1])$) In addition a partial classification has been given in the case

A is an AF-algebra

The knowledge of classifications of general closed *-derivations is therefore very rudimentary, and as Theorem 1.6.20 and 21 shows, the solution of the problem in other cases will be extremely complicated if it can be found at all. However, if the domain $D(\delta)$ is specified in advance as a dense *subalgebra $\mathcal{D}$ of A, the classification is often quite simple. If for example $A = \overline{\bigcup_n A_n}$ is an AF-algebra, and $\mathcal{D} = \bigcup_n A_n$ (without closure), the solution is given by Remark 1.6.7. If X is a differentiable compact manifold, $A = C(X)$, and $\mathcal{D} = C^\infty(X)$ is the subalgebra of infinitely many times differentiable functions on X, then the derivations $\delta : \mathcal{D} \to A$ are nothing but the vector fields on X, and it is well known that δ has the form

$$\delta = \sum_{k=1}^{d} f_k(x) \frac{\partial}{\partial x_k}$$

in the local coordinate charts of X, where $d = \text{dimension}(X)$, $\frac{\partial}{\partial x_k}$ is the derivatives in the d coordinate directions and f_k are continuous functions on the local charts.

Note that the topological space X is uniquely determined from the algebra A as the space of maximal ideals, while the differentiable structure of X is determined by $\mathcal{D} = C^{\infty}(X)$, i.e. the pair $(A, \mathcal{D})$ determines the differentiable manifold X. Thus a non-commutative manifold could be defined as a pair $(A, \mathcal{D})$ consisting of a C*-algebra A and a dense *-subalgebra $\mathcal{D}$, where $\mathcal{D}$ should satisfy some additional properties. What these properties are seems at present not very clear; one could roughly envisage a noncommutative compact manifold as (A, A_{∞}), where A_{∞} is the set of elements in contained in the domain of all polynomials of d given closed derivations $\delta_1, \ldots, \delta_d$ on A with the property that $(1 - \sum_{k=1}^{d} \delta_k^2)^{-1}$ exists and is compact (This is (my understanding of) a suggestion of Connes, which was conveyed to me by Cuntz). The compactness assumption corresponds to the compactness of the resolvent of the Laplacian on an ordinary compact manifold, and it excludes for example the case that $A = C(X)$ where X is a compact Hausdorff space and $\delta_1, \ldots, \delta_d$ are the generators of a flow of $\mathbb{R}^d$ on X which is ergodic, but not transitive, but it includes the case that the action is transitive (and then $X = \mathbb{T}^n$ = the n-dimensional torus, where $n < d$).

In these notes we will not attempt to make a general definition of non-commutative manifolds, but only consider the case that $\mathcal{D}$ consists of the smooth elements of the action α of a locally compact group G on a C*-algebra A as a strongly continuous group of *-automorphisms. In the case that A is abelian this does not neccesarily mean that the spectrum X has a manifold structure, but has a "foliation" structure where the leaves are the G-orbits, and the requirement that $D(\delta)$ is the smooth elements means that δ is tangential to the G-orbits in a sense which will be made precise in Theorem 2.4.11. The classes of smooth elements we will consider are

A_n = the *-algebra of n times differentiable elements with respect to the one-parameter subgroups of $\alpha(G)$; $n = 1,2,3,\ldots,\infty$.

A_F = the algebra of G-finite elements

$$= \begin{cases} \text{the set of elements with compact spectrum in the dual group } \hat{G}, \text{ if } G \text{ is abelian} \\ \text{the set of elements } x \text{ such that the linear span of } \alpha_G(x) = \{\alpha_g(x) \mid g \in G\} \text{ is finite dimensional, if } G \text{ is compact.} \end{cases}$$

The precise definitions will be made in section 2.2. One has

$$A_F \subseteq A_\infty \subseteq \cdots \subseteq A_2 \subseteq A_1 \subseteq A$$

and all these *-algebras are dense in A. The general problem is then to classify all *-derivations mapping one class of smooth elements into another.

Definition 2.1.1

If $\mathcal{D}_1$, $\mathcal{D}_2$ are dense *-subalgebras of a C*-algebra A, $\mathrm{Der}(\mathcal{D}_1, \mathcal{D}_2)$ denotes the space of all *derivations δ with $D(\delta) = \mathcal{D}_1$ and $R(\delta) \equiv \delta(\mathcal{D}_1) \subseteq \mathcal{D}_2$. A derivation $\delta \in \mathrm{Der}(\mathcal{D}_1, \mathcal{D}_2)$ is said to be approximately inner if there exists a net $h_n \in A$, (which can be taken to consist of skew-adjoint elements) such that

$$\lim_n \| \delta(x) - (h_n x - x h_n) \| = 0$$

for all $x \in \mathcal{D}_1$.

Example 2.1.2

If A is separable, all derivations in $\mathrm{Der}(A, A)$ are approximately inner, see [Ell 2], Remark 3.2. (See also Theorem 1.2.3.) If $A = \overline{\bigcup_n A_n}$ is an AF-algebra, then all derivations in $\mathrm{Der}(\bigcup_n A_n, A)$ are approximately inner, see Remark 1.6.7.

Problem 2.1.3

When does all derivations in $\mathrm{Der}(A_F, A)$ (or some other class) have a decomposition

$$\delta = a_1\delta_1 + a_2\delta_2 + \cdots + a_d\delta_d + \tilde{\delta},$$

where

$\delta_1, \ldots, \delta_d$ is a basis for the action of the Lie algebra of G on A(i.e. $\delta_1, \ldots, \delta_d$ span linearly all the derivatives of one-parameter subgroups of $\alpha(G)$, if this linear span is infinite dimensional the question must be reformulated);

$a_1, a_2, \ldots, a_d$ are real functions on the primitive ideal spectrum $\mathrm{Prim}(A)$ of A;

$\tilde{\delta}$ is an approximately inner, or inner, or bounded (depending on context) derivation;
and the identity above means

$$\pi(\delta(x)) = a_1(\ker \pi)\pi(\delta_1(x)) + \ldots + a_d(\ker \pi)\pi(\delta_d(x)) + \pi(\tilde{\delta}(x))$$

for all $x \in A_F$ and all irreducible representations π of A. Thus if A is primitive (i.e. A has a faithful irreducible representation), $a_1, \ldots, a_d$ are real scalars.

Problem 2.1.3 is related to the question of Sakai in [Kad 1], which induced the theory of noncommutative vectorfields.

Positive answers to the above question are given in Theorem 2.4.1' Theorem 2.4.15, Theorem 2.4.20, Theorem 2.6.6, Example 2.7.10, Theorem 2.9.1, Remark 2.9.2, Theorem 2.9.3, Corollary 2.9.17, Theorem 2.9.22, Corollary 2.9.26, Theorem 2.9.31, Theorem 2.9.42, Theorem 2.9.43 and Theorem 2.9.45.
Note in particular that Theorem 2.9.3, which treats a situation close to the noncommutative manifold situation mentioned above, gives a particularly clear answer to the question. There are also negative answers to Problem 2.1.3., even in the case that A is simple with identity and G is the circle group $\mathbb{T}$, see example 2.7.10. This leads to the problem of computing the quotient $H^1(A_F, A)$ of Der (A_F, A) modulo the approximately inner derivations. H^1 is the first cyclic cohomology-group in Connes's theory, [Con 1] (or at least related to this group, it is actually the first Hochschild cohomology group).

<u>Problem 2.1.4</u>

If G is a Lie group, are all derivations in Der (A_F, A_F) pregenerators? Same question can be asked for Der(A_∞, A_1).

The answer of this question is not known in general, but we prove special cases in Theorem 2.4.24, Theorem 2.4.26, Theorem 2.4.28, Theorem 2.4.30, Theorem 2.6.1, Theorem 2.6.2, Theorem 2.6.8, Theorem 2.9.7, Theorem 2.9.10, Theorem 2.9.18, Theorem 2.9.24, Theorem 2.9.28, Theorem 2.9.40, Theorem 2.9.42, Theorem 2.10.1 and Theorem 2.10.2. It is known that the answer is false, even for Der(A_∞, A_∞), when G is not a Lie group, see Example 2.9.21.

In connection with actions α of locally compact groups G on C^*-algebras it is also natural (at least when G is abelian) to consider closed derivations δ commuting with the action in the sense that $D(\delta)$ is α-invariant and

$$\delta\alpha_g(x) = \alpha_g\delta(x)$$

for all $x \in D(\delta)$, or in short

$$[\delta, \alpha] = 0$$

In this case the condition that $D(\delta)$ is a class of smooth elements is often not crucial, because regularization techniques allows us to conclude that $D(\delta)$ contains a core for δ consisting of smooth elements. In this case δ restricts to a derivation of the fixed point algebra A^α for the action α, and when G is compact several generator results have been proved when $\delta|_{A^\alpha}$ satisfy various conditions, like $\delta|_{A^\alpha} = 0$ or $A^\alpha \subseteq D(\delta)$ or $\delta|_{A^\alpha}$ is a generator. Some of these results will be studied in Section 2.8. Another condition on δ which has been considered is almost commutation of δ with α in the sense that $[\delta, \alpha_G]$ has a finite-dimensional linear span, see Section 2.10. Also, when G is not compact, the condition $\delta|_{A^\alpha} = 0$ has been replaced by various stronger conditions of "locality" of δ with respect to the action, see Remark 2.4.12 and [BaR 2], [Bat 7], [BDE 1], [BDR 1], [BEE 1], [BER 1], [BER 3], [BER 4], [Bra 5], [Rob 4].

When δ is a complete dissipation, classification and generator results have only been obtained when δ is bounded, see Chapter 3.1, or when G is compact and abelian, $[\delta, \alpha] = 0$, $A_F^\alpha \subseteq D(\delta)$ and $\delta|_{A^\alpha} = 0$. At present it is not even clear how the analogue of Problem 2.1.3 should be formulated in general.

§2.2 Classes of smooth elements and spectral theory for group actions

In this section we introduce the classes $A_F \subseteq A_\infty \subseteq \ldots \subseteq A_n \subseteq A_{n-1} \subseteq \ldots \subseteq A_1 \subseteq A$ of smooth elements alluded to in Section 2.1.

§2.2.1 Differentiable elements

We use the following notations.
A: a Banach space, G: a locally compact group.
α_g: a strongly continous action of G on A as isometries.

(2.2.1) $\Delta \equiv \{\frac{d}{dt}\alpha_{\theta(t)}\,|_{t=0} \mid t \to \theta(t)$ continuous morphism from $\mathbb{R}$ to $G\}$
= the set of infinitesimal generators of the actions of the one-parameter subgroups of G on A.

(2.2.2) $A_n \equiv \{x \in A \mid x \in D(\delta_1\delta_2 \ldots \delta_n)$ for all $\delta_1, \ldots, \delta_n \in \Delta\}$
$\equiv$ the n times differentiable elements in A,
$n = 1, 2, 3, \ldots.$

(2.2.3) $A_\infty \equiv \bigcap_{n=1} A_n \equiv$ the infinitely many times differentiable elements in A.

Proposition 2.2.1

If G is a Lie group, or if G is compact, or if G is abelian, then A_∞ is dense in A.

Proof

If G is a Lie group, and $f \in C^\infty_{00}(G)$ = the C^∞-functions on G with compact supports, then $\alpha_f(x) \in A_\infty$ for all $x \in A$, see (1.5.6) and (1.5.5). Letting f run through a sequence converging to δ weakly, we can approximate $x \in A$ with $\alpha_f(x) \in A_\infty$.

If G is a compact group, G is a projective limit of compact Lie groups since the irreducible unitary representations of G are finite dimensional (and thus with the range in a closed subgroup of some $U(n)$, i.e. in a Lie group) and separates points. Put in another way, there is a decreasing net H_τ of closed normal subgroups of G such that $\bigcap_\tau H_\tau = \{e\}$ and G/H_τ is Lie for each τ. Now, if $x \in A$ we can approximate x in norm with elements of the form $x_\tau = \int_{H_\tau} \alpha_h(x)dh$, where dh is Haar measure on H_τ. On the G-invariant subspace $A_\tau = \int_{H_\tau} \alpha_h(A)dh$, the action of G lifts to an action of the Lie group G/H_τ, and hence each x_τ can be approximated by elements in $A_{\tau,\infty} \subseteq A_\infty$ by the first part of the proof. Thus A_∞ is dense in .

If G is abelian, we will show in Section 2.2.4 that A_F(= the set of G finite elements) is contained in A_∞, and A_F is dense, see Proposition 2.2.14.

Problem 2.2.2

Is A_∞ dense in A for a general locally compact group G ?

A_n is a locally convex space with the topology induced by the seminorms $\|\delta_1 \cdots \delta_m x\|$ ($x \in A$, $\delta_1, \ldots, \delta_m \in \Delta$ and $m < n + 1$.). If G is a Lie group, Δ is finite dimensional, and then A_n is a Banach space for $n < \infty$, while A_∞ is a Fréchet space.

Above we constructed elements in A_∞ by regularizing with $f \in C_{00}^\infty(G)$. It is a deep result that any element in A_∞ can be obtained in this manner:

Theorem 2.2.3 (Dixmier-Malliavin)

If G is a Lie group, α is a action of G on a Banach space A and V is a neighbourhood of $e \in G$, then any x in A_∞ is a finite linear combination of the form

$$(2.2.4) \quad x = \sum_{\ell=1}^{n} \alpha_{f_\ell}(x_\ell) \quad \text{where } x_\ell \in A_\infty, \; f \in C_{00}^\infty(G) \text{ and } \operatorname{supp} f_\ell \subset V .$$

For the proof see [DM 1], Théorème 3.3. It is interesting that for a wide class of groups, including semi-finite groups with finite center, the n in the above theorem can be taken to be 1, while for other groups, like $G = \mathbb{R}^2$, n in general must be chosen greater than 1.

Example 2.2.4

Let A be a C* algebra, α a group of *automorphisms. Then Δ consists of derivations and A_n is a *algebra. The derivation property implies that

$$(2.2.5)\quad \delta_1\delta_2 \ldots \delta_k(xy) = \sum_P \delta_P(x)\, \delta_{\bar{P}}(y)$$

where P runs over all subsets of $\{1,2,\ldots k\}$, $\bar{P} = \{1,2,\ldots,k\} \setminus P$, and

$$(2.2.6)\quad \delta_P = \delta_{i_1}\delta_{i_2} \ldots \delta_{i_s} \quad \text{if} \quad P = \{i_1, \ldots, i_s\} \text{ with } i_1 < \ldots < i_s.$$

It follows that the product operation is continuous, i.e. A_n is a topological *algebra. If $n < p$ and Δ is finite dimensional, $\Delta = \text{span}\{\delta_1 \ldots \delta_d\}$, then A_n is a Banach algebra with the norm

$$(2.2.7)\quad \|x\|_n = \|x\| + \sum_{k=1}^{n} \sum_{i_1=1}^{d} \ldots \sum_{i_k=1}^{d} \|\delta_{i_I} \ldots \delta_{i_k} x\| / k!$$

To show $\|xy\|_n \leq \|x\|_n \|y\|_n$, use the above formula.

§2.2.2 Spectral theory for abelian groups

Let G be a locally compact abelian group with Haar measure dt and $\hat{G}$ be the dual group of G, i.e. the set of all unitary characters with the dual topology, [Rud 1]. For $\gamma \in \hat{G}$, $t \in G$ $\gamma(t) \in \mathbb{T} = \{z \in \mathbb{C} \mid |z| = 1\}$. Let $L^1(G)$ be the group algebra of G. For $f,g \in L^1(G)$, the product and involution is defined by

$$(2.2.8)\quad f*g(t) = \int_G f(t - s)g(s)ds$$

$$(2.2.9)\quad f^*(t) = \overline{f(-t)}$$

The Fourier transform is defined by

$$(2.2.10)\quad \hat{f}(\gamma) = \int_G \gamma(t)\, f(t)dt \ ; \quad f \in L^1(G) \to \hat{f} \in C_0(\hat{G})$$

This is an algebra-morphism with dense range in $C_0(\hat{G})$.
Let $t \to \alpha_t$ be an isometric action of G on a Banach space A.

Definition 2.2.5

Let Y be a subset of A. Recall that $\alpha_f(x) = \int \alpha_t(x)f(t)dt$ for $x \in A$, $f \in L^1(G)$. Define

$$(2.2.11) \quad I_Y^\alpha \equiv \{f \in L^1(G) \mid \alpha_f(x) = 0 \text{ for all } x \in Y\}.$$

Since $\alpha_{f*g} = \alpha_f \cdot \alpha_g$, I_Y^α is an ideal of $L^1(G)$. The α-Spectrum of Y, $\mathrm{Spec}^\alpha(Y) = \mathrm{Spec}\,(Y)$ is defined as the hull of this ideal, i.e.

$$(2.2.12) \quad \mathrm{Spec}^\alpha(Y) \equiv \{\gamma \in \hat{G} \mid \hat{f}(\gamma) = 0 \text{ for all } f \in I_Y^\alpha\}.$$

The Spectrum of α is defined by

$$(2.2.13) \quad \mathrm{Spec}\,(\alpha) = \mathrm{Spec}^\alpha(A).$$

For a subset E of $\hat{G}$, the spectral subspaces $A(E)$ and $A_0^\alpha(E)$ are defined by

$$(2.2.14) \quad A^\alpha(E) \equiv \overline{\{x \in A \mid \mathrm{Spec}^\alpha(x) \subset E\}}^{\text{norm}}$$

and

$$(2.2.15) \quad A_0^\alpha(E) \equiv \text{the closed linear span of } \{\alpha_f(x) \mid \mathrm{supp}\,\hat{f} \subseteq E,\ x \in A\}$$

Example 2.2.6

Assume $x (\in A)$ satisfies the equation $\alpha_t(x) = \gamma(t)x$ for a $\gamma \in \hat{G}$, and all $t \in G$. Then for $f \in L^1(G)$, $\alpha_f(x) = \hat{f}(\gamma)x$, hence $I_{\{x\}}^\alpha = \{f \in L^1(G) \mid \hat{f}(\gamma) = 0\}$ and thus $\mathrm{Spec}^\alpha(x) = \{\gamma\}$. By the Tauberian theorem, the converse is also true. So $\mathrm{Spec}^\alpha(x) = \{\gamma\} \Leftrightarrow \alpha_t(x) = \gamma(t)x$ for all $t \in G$.

The following properties are easy to check, see [BR1], Lemma 3.2.38.

Proposition 2.2.7

$$(2.2.16) \quad \mathrm{Spec}^\alpha(\alpha_t(x)) = \mathrm{Spec}^\alpha(x)$$

(2.2.17) $\mathrm{Spec}^{\alpha}(x + y) \subseteq \mathrm{Spec}^{\alpha}(x) \cup \mathrm{Spec}^{\alpha}(y)$

(2.2.18) $\mathrm{Spec}^{\alpha}(\alpha_f(x)) \subseteq \mathrm{supp}\,\hat{f} \cap \mathrm{Spec}^{\alpha}(x)$.

(2.2.19) If $f_1, f_2 \in L^1(G)$ and $\hat{f}_1 = \hat{f}_2$ in a neighbourhood of $\mathrm{Spec}^{\alpha}(x)$, then $\alpha_{f_1}(x) = \alpha_{f_2}(x)$.

By these properties, we have

Proposition 2.2.8

(2.2.20) $A_0^{\alpha}(E) \subseteq A^{\alpha}(E)$

(2.2.21) $\alpha_t(A^{\alpha}(E)) = A^{\alpha}(E)$.

(2.2.22) If E is closed, $A^{\alpha}(E) = \{x \mid \mathrm{Spec}\ x \in E\}$, i.e. the set of the right hand side is automatically closed.

(2.2.23) If E is open, $A^{\alpha}(E) = A_0^{\alpha}(E)$
$= \overline{\cup\{A^{\alpha}(K) \mid K \text{ compact subset of } E\}}^{\text{norm}}$

(2.2.24) If E is closed, $A^{\alpha}(E) = \bigcap_V A_0^{\alpha}(E + V)$ V neighbourhood of the identity of $\hat{G}$.

See [BR1], Lemma 3.2.39, for details.

Example 2.2.9

Assume that A is a C*-algebra and α is a representation of G into the *automorphisms of A.
Suppose $x, y \in A$ satisfy the equations $\alpha_t(x) = \gamma_1(t)x, \alpha_t(y) = \gamma_2(t)y$,

then $\alpha_t(xy) = (\gamma_1 + \gamma_2)(t)xy$ $\alpha_t(x^*) = (-\gamma_1)(t)x^*$.
More generally the following are true.

(2.2.25) $\mathrm{Spec}^{\alpha}(x^*) = -\mathrm{Spec}^{\alpha}(x)$.

(2.2.26) $A^{\alpha}(E)^* = A^{\alpha}(-E)$.

(2.2.27) $A_0^{\alpha}(E_1)\,A_0^{\alpha}(E_2) \subseteq A_0^{\alpha}(E_1 + E_2)$ when E_1 and E_2 are open.

(2.2.28) $A^{\alpha}(E_1)\, A^{\alpha}(E_2) \subseteq A^{\alpha}(\overline{E_1+E_2})$.

(2.2.29) $\mathrm{Spec}^{\alpha}(xy) \subseteq \overline{\mathrm{Spec}^{\alpha}(x) + \mathrm{Spec}^{\alpha}(y)}$.

As a special case of (2.2.28) and (2.2.26).

(2.2.30) $A^{\alpha}(\gamma_1)\, A^{\alpha}(\gamma_2) \leq A^{\alpha}(\gamma_1+\gamma_2)$ $\quad A^{\alpha}(\gamma)^* = A^{\alpha}(-\gamma)$.

Note that

$A^{\alpha} \equiv A^{\alpha}(0) = \{x \in A \mid \alpha_t(x) = x\}$ = the fixed point algebra under the action α .

§2.2.3. Spectral theory for compact groups

Definition 2.2.10

If G is compact abelian and the Haar measure dt is normalized, set

(2.2.31) $P_{\gamma}(x) \equiv \int_G \overline{\gamma}(t)\, \alpha_t(x)\, dt \qquad (x \in A)$.

Then P_{γ} is a projection of norm 1 from A onto $A^{\alpha}(\gamma)$, and we have the formulas

(2.2.32) $\alpha_t(P_{\gamma}x) = \gamma(t) P_{\gamma}(x)$

and

(2.2.33) $x = \sum_{\gamma} P_{\gamma}(x)$

for $x \in A$, where the last expansion is convergent in Cesaro mean. Thus any element $x \in A$ has a spectral decomposition with respect to the action α.
If G is compact, but nonabelian, we can define a corresponding decomposition. So, let G be a compact group with the normalized Haar measure dg, α an isometric representation of G on a Banach space A, $\hat{G}$ = the set of all equivalence classes of ire-ducible representations of G.

For $\gamma \in \hat{G}$, put $d(\gamma)$ = the dimension of γ, and fix a matrix representative $\gamma(g) = [\gamma_{ij}(g)]_{i,j=1}^{d(\gamma)}$ of γ.

Definition 2.2.11

(2.2.34) $$P_{ij}(\gamma)(x) = \int_G d(\gamma)\, \overline{\gamma_{ji}(g)}\, \alpha_g(x)\, dg,$$

and

(2.2.35) $$P_\gamma(x) = \sum_{i=1}^{d(\gamma)} P_{ii}(\gamma)(x)$$

$$= \int_G d(\gamma)\, \overline{\mathrm{tr}\,(\gamma(g))}\, \alpha_g(x) dg$$

for $x \in A$.

Using Weyl's orthogonality relations for the matrix elements $\gamma_{ij}(y)$, see [HR 2] , Theorem 27.19, one successively verifies that

(2.2.36) $$P_{ij}(\gamma)P_{k\ell}(\gamma) = \delta_{i\ell}P_{kj}(\gamma),$$

(2.2.37) $$P_\gamma P_{ij}(\gamma) = P_{ij}(\gamma)P_\gamma = P_{ij}(\gamma),$$

(2.2.38) $$P_\gamma^2 = P_\gamma ,$$

and using that γ is a unitary representation one has the transformation law

(2.2.39) $$\alpha_g(P_{ij}(\gamma)(x)) = \sum_{k=1}^{d(\gamma)} P_{ik}(\gamma)(x)\gamma_{kj}(g).$$

analogous to (2.2.32).
Furthermore one has the completeness relation

(2.2.40) $$x = \sum_{\gamma \in \hat{G}} P_\gamma(x),$$

analogous to (2.2.33), where the sum again converges in a suitable mean, and this gives the desired spectral decomposition. We see that there are two suitable analogues of the spectral subspaces $A^\alpha(\gamma)$ for abelian G in this case:

Definition 2.2.12

Define

(2.2.41) $\quad A^{\alpha}(\gamma) = P_{\gamma}(A)$

and if n is a positive integer, define $A^{\alpha}_{n}(\gamma)$ as the set of $n \times d(\gamma)$ matrices $[x_{ij}]$ with entries $x_{ij} \in A$ satisfying the transformation law

(2.2.42) $\quad \alpha_g([x_{ij}]) = [x_{ij}]\,(1\!\!1_A \otimes \gamma(g))$

or, equivalently

(2.2.43) $\quad \alpha_g(x_{ij}) = \sum_k x_{ik}\gamma_{kj}(g)$

In particular, for $x \in A$, it follows from 2.2.39 that

(2.2.44) $\quad [P_{ij}(\gamma)(x)] \in A^{\alpha}_{d(\gamma)}(\gamma).$

§2.2.4 The algebra of G-finite elements

Definition 2.2.13

Let G be an abelian, locally compact group, and α be an action of G on C*-algebra A. The set

(2.2.45) $\quad A^{\alpha}_{F} = \{x \in A \;;\; \mathrm{Spec}^{\alpha}(x) \text{ is compact in } \hat{G}\}$

is called the algebra of G-finite elements. Then

(2.2.46) $\quad A^{\alpha}_{F} = \bigcup\{A^{\alpha}(K);\ K \text{ is a compact subset of } \hat{G}\}$.

A^{α}_{F} is a *-subalgebra of A, because of the relations

(2.2.47) $\quad \begin{cases} A^{\alpha}(K_1)\,A^{\alpha}(K_2) \subseteq A^{\alpha}(K_1 + K_2) , \\ A^{\alpha}(K)^* = A^{\alpha}(-K) , \end{cases}$

see Example 2.2.9.

Proposition 2.2.14

A_F^α is dense in A, and $A_F^\alpha \subseteq A_\infty$.

Proof

The relation (2.2.23), applied to $E = \hat{G}$, implies that A_F^α is dense in A. For the second statement it suffices to show that $A^\alpha(K) \subseteq A_\infty$ for all compact subsets K of $\hat{G}$. But this follows from the fact that $A^\alpha(K)$ is globally α-invariant and that $\alpha|_{A^\alpha(K)}$ is norm-continuous. Let's prove the norm-continuity. Assume that $f \in L^1(G)$ satisfies $\hat{f} = 1$ in a neighbourhood of K. Then $x = \alpha_f(x)$ for all $x \in A^\alpha(K)$. Hence $\| \alpha_g(x) - x\| = \| \alpha_{f(\cdot-g)}(x) - \alpha_f(x)\| \leqq \|f(\cdot-g) - f\|_1 \|x\|$. Therefore, $\| (\alpha_g - 1)|_{A^\alpha(K)}\| \le \| f(\cdot-g) - f\|_1 \to 0 \quad (g \to 0)$.

If G is compact and abelian, then

(2.2.48) A_F^α = the linear span of $\{ A^\alpha(\gamma); \ \gamma \in \hat{G}\} = \{x \in A$; the linear span of the orbit $\{\alpha_g(x);\ g \in G\}$ is finite-dimensional$\}$.

Definition 2.2.15

If G is compact (G may be non-abelian), use 2.2.48 as definition of A_F^α. Then A_F^α is still a dense *-subalgebra of A and $A_F^\alpha \subseteq A_\infty$, by 2.2.40 and similar reasoning as in Proposition 2.2.14.

Example 2.2.16

Let G be a compact abelian group, let $A = C(G)$ and let α be the action by translation, $(\alpha_g f)(h) = f(g + h)$. If $\gamma \in \hat{G}$ and $f \in A^\alpha(\gamma)$, then $f(t) = (\alpha_t f)(0) = \gamma(t)f(0)$ for all $t \in G$, hence $f \in \mathbb{C}\gamma$, and since conversely $f = \gamma$ has the transformation property above, we deduce that $A^\alpha(\gamma)$ is the one-dimensional space spanned by γ. It follows that A_F^α is the space of trigonometric polynomials on G.

§2.3 Closability and automatic continuity of non-commutative vector fields.

§2.3.1 Automatic continuity of derivations of A_n, $n=1,2,\cdots,\infty$.

Theorem 2.3.1 (Johnson, Ringrose, Longo, Bratteli-Elliott-Jørgensen)

Let Δ be a finite or countable collection of closed, densely defined *-derivations on a C*-algebra A, and define

(2.3.1) $$A_n = \bigcap\{D(\delta_1\cdots\delta_k \mid k<n+1,\ \delta_i\in\Delta,\ i=1,\cdots,k\}$$

where n is a positive integer or $n=\infty$. Assume that A_n is dense in A. If $\delta\in \mathrm{Der}\,(A_n, A)$, then there exists finitely many elements $\delta_1,\cdots,\delta_m \in\Delta$, a constant $C>0$, and a finite $n_0\leq n$ such that

(2.3.2) $$\|\delta(x)\| \leq C(\|x\| + \sum_{k=1}^{n_0}\sum_{i_1=1}^{m}\cdots\sum_{i_k=1}^{m}\|\delta_{i_1}\cdots\delta_{i_k}x\|/k!)$$

for all $x\in A_n$.

Remark 2.3.2

We will follow the argument in [BEJ 1], but the ideas of the proof were developed successively in [Joh 1], [Rin 1] and [Lon 1]. The implicit statement in [BEJ 1] that density of A_n in A is not needed in the proof is erroneous, it is used in the final step : $K=\{0\}\Rightarrow \delta : A_n \to A$ is closed, see observation 3 below.

We need the following refinement of Theorem 1.6.2.

Lemma 2.3.2

If A has an identity, $x=x^*\in A_n$, and f is a $n+1$ times continuously differentiable function on $\mathbb{R}$, then $f(x)\in A_n$.

Proof

It is enough to show this for finite n, and by modifying f outside the spectrum of x we may assume that f has compact support. Then the function $p\to |p|^k|\hat{f}(p)|$ is integrable for $k=0,1,\cdots,n$. By Theorem 1.6.2, $f(x)\in D(\delta_1)$ for any $\delta_1\in\Delta$, and

$$(2.3.3) \qquad \delta_1(f(x)) = i(2\pi)^{-\frac{1}{2}} \int_{-\infty}^{\infty} p\hat{f}(p)\left(\int_0^1 e^{itpx}\delta_1(x)e^{i(1-t)px}dt\right)dp$$

Also, by (1.6.12), if $\delta_2 \in \Delta$, $e^{itpx} \in D(\delta_2)$ and

$$(2.3.4) \qquad \delta_2(e^{itpx}) = itp\int_0^1 e^{istpx}\ \delta(x)e^{i(1-s)tpx}\ ds$$

As δ_2 is closed, it follows from the derivation property of δ_2 and (2.3.3) that $\delta_1(f(x)) \in D(\delta_2)$ and

$$\delta_2\delta_1(f(x)) = -(2\pi)^{-\frac{1}{2}}\int_{-\infty}^{\infty} dp\ p^2\hat{f}(p)\int_0^1 dt\int_0^1 ds$$

$$(2.3.5) \qquad [t\,e^{istpx}\ \delta_2(x)\,e^{i(1-s)tpx}\ \delta_1(x)\,e^{i(1-t)px}$$

$$+ (1-t)\ e^{itpx}\ \delta_1(x)\ e^{is(1-t)px}\ \delta_2(x)\ e^{i(1-s)(1-t)px}]$$

$$+ i(2\pi)^{-\frac{1}{2}}\int_{-\infty}^{\infty} dp\,p\,\hat{f}(p)\int_0^1 dt\ e^{itpx}\ \delta_2\delta_1(x)\ e^{i(1-t)px}.$$

By repeating this argument, it follows that $f(x) \in D(\delta_n \cdots \delta_1)$ for all $\delta_1, \cdots, \delta_n \in \Delta$, i.e. $f(x) \in A_n$.

Proof of Theorem 2.3.1

By step 2 of Theorem 1.6.2, we may assume that A has an identity $\mathbb{1}$. Put

$$I = \{y \in A_n \mid x \in A_n \to \delta(xy) \text{ is continuous}\}$$

Here by continuity of $x \to \delta(xy)$ we mean continuity as a map from A_n in its Fréchet topology into A with norm topology. The derivation property of δ implies that I is a twosided ideal in A_n. Set $J = A_n \cap \bar{I}$, where $\bar{I}$ is the closure of I in A. Then J is a closed ideal in A_n.

Observation 1

A_n/J is finite dimensional.

Proof

A_n/J identifies with the canonical image of A_n in the C*-algebra $A/\bar{I}$ under the quotient map ϕ, and thus A_n/J can be represented as a *-algebra of operators on a Hilbert space.

Assume ad absurdum that A_n/J is infinite dimensional. We first argue that A_n/J contains a self-adjoint element h with infinite spectrum. Assume conversely that all selfadjoint elements in A_n/J has finite spectrum. Then all selfadjoint elements in A_n/J is a finite linear combination of mutually orthogonal projections, in particular $A_n/J = \mathcal{B}$ contains a projection p such that $0 \neq p \neq \mathbb{1}$. Now, at least one of the algebras $p\mathcal{B}p$ or $(\mathbb{1}-p)\mathcal{B}(\mathbb{1}-p)$ must be infinite dimensional, because a simple argument, squeezing $\mathcal{B}$ between minimal partial isometries (i.e. matrix elements) in $p\mathcal{B}p$ and $(\mathbb{1}-p)\mathcal{B}(\mathbb{1}-p)$ shows that

$$\text{dimension } (p\mathcal{B}(\mathbb{1}-p)) \leq \dim (p\mathcal{B}p)\dim((\mathbb{1}-p)\mathcal{B}(\mathbb{1}-p))$$

while

$$\dim ((\mathbb{1}-p)\,\mathcal{B}\,p) = \dim (p\mathcal{B}(\mathbb{1}-p))$$

and $\mathcal{B}$ is the direct sum of the four subspaces $p\mathcal{B}p$, $p\mathcal{B}(\mathbb{1}-p)$, $(\mathbb{1}-p)\mathcal{B}p$ and $(\mathbb{1}-p)\mathcal{B}(\mathbb{1}-p)$. If $p\mathcal{B}p$ is infinite-dimensional, put $p_1 = \mathbb{1}-p$, if not, put $p_1 = p$. Repeating this argument, with $\mathbb{1}$ replaced by $\mathbb{1}-p$, we find a nonzero projection $p_2 \leq \mathbb{1}-p_1$ such that $(\mathbb{1}-p_1-p_2)\mathcal{B}(\mathbb{1}-p_1-p_2)$ is infinite dimensional, and by iteration we get a sequence p_k of nonzero, mutually orthogonal projections in $\mathcal{B}$. Let x_k be selfadjoint elements in A_n such that $\phi(x_k) = p_k$. As Δ is countable, the topology on A_n is given by an increasing sequence of algebra norms

$$\|\cdot\|_1 = \|\cdot\| \leq \|\cdot\|_2 \leq \|\cdot\|_3 \leq \cdots .$$

For each k, choose inductively a number λ_k such that $0 < \lambda_k < \lambda_{k-1}$ and

$$\lambda_k \|x_k\|_k \leq 2^{-k}$$

Then $h = \sum_k \lambda_k x_k$ converge in the topology of A_n to a selfadjoint element in A_n, and

$$\text{Spec } (\phi(h)) = \text{Spec } (\sum_k \lambda_k p_k) = \{\lambda_k \mid k = 1, 2, \ldots\} \cup \{0\}$$

Thus $\phi(h)$ has infinite spectrum, and this contradiction establishes that there is a selfadjoint operator $h \in A_n$ such that $\phi(h) \in A_n/J$ has infinite spectrum.

If f is $n+1$ times continuously differentiable, then $f(h) \in A_n$ by Lemma 2.3.2. Choosing a sequence of f's in $C^{n+1}(\mathbb{R})$ with disjoint supports such that each f is nonzero on Spec $(\phi(h))$, we obtain a sequence $h_m = h_m^* \in A_n$ such that

$$h_m h_k = 0 \quad \text{for} \quad m \neq k , \qquad h_m^2 \notin J$$

Normalize h_m so that

$$\| h_m \|_m = 1 , \qquad m = 1,2,\cdots .$$

Since $h_m^2 \notin I (\subseteq J)$, there is a sequence x_n in A_n such that

$$\| x_m \|_m < 2^{-m} \quad \text{and} \quad \| \delta(x_m h_m^2) \| > \| \delta(h_m) \| + m$$

Put

$$z = \sum_m x_m h_m$$

If $m \geq k$, then

$$\| x_m h_m \|_k \leq \| x_m \|_k \| h_m \|_k \leq \| x_m \|_m \| h_m \|_m \leq 2^{-m}$$

and it follows that the sum defining z converges in the topology of A_n, in particular $z \in A_n$. Also $\| z \| \leq \sum_m 2^{-m} = 1$ and $zh_m = x_m h_m^2$ and hence

$$\begin{aligned} \| \delta(z) h_m \| &= \| \delta(zh_m) - z\, \delta(h_m) \| \\ &\geq \| \delta(x_m h_m^2) \| - \| z \| \| \delta(h_m) \| \geq m \end{aligned}$$

As $\| h_m \| \leq 1$, this contradicts $z \in D(\delta)$, and Observation 1 follows.

As A_n is dense in A, this implies that A/I is finite-dimensional. Put

$$K = \{x \in A \mid xy = 0 \quad \text{for all} \quad y \in \bar{I}\}$$

<u>Observation 2.</u>

$K = \{0\}$.

<u>Proof</u>

As I is a twosided ideal in A, so is K.
As $K\bar{I} = \{0\}$, it follows that $K \cap \bar{I} = \{0\}$, and hence K is finite dimensional. Thus K has an identity p, and p is a central projection in A with $K = Ap = pA$.

By Lemma 2.3.2. and the argument used in Example 1.6.4. there is a projection $q \in A_n$ such that $\|q-p\| < 1$, but as p is central, it follows that $q = p$, and hence $p \in A_n$. Since K is finite-dimensional, the map $x \in A_n \to xp \in K \cap A_n \to \delta(xp) \in A$ is continuous. Thus $p \in K \cap I = \{0\}$ and this implies $K = A_p = \{0\}$.

<u>Observation 3</u>

The map $\delta : A_n \to A$ is closed when A_n is equipped with its Fréchet topology.

<u>Proof</u>

If x_m is a sequence in A_n such that

$$\|x_m\|_m \underset{n\to\infty}{\to} 0 \text{ for all } k, \quad \delta(x_m) \to z$$

then $\delta(x_m y) \to 0$ for all $y \in I$, and

$$\delta(x_m y) = \delta(x_m)y + x_m\,\delta(y) \to zy$$

for all $y \in I$. Thus $z \in K = \{0\}$. This proves that δ is closed.

The topology on A_n is determined by the complete metric

$$d(x,y) = \sum_m 2^{-m} \|x-y\|_m (1+\|x-y\|_m)^{-1}$$

Therefore A_n is of the second category, and the closed graph theorem applies to show that δ is continuous [KN 1], 2.11.3. It follows that δ is bounded as stated in (2.3.2)

Problèm 2.3.3

Adopt the hypothesesof Theorem 3.2.1 but assume in addition that A is primitive. Does it follow that any $\delta \in \mathrm{Der}(A_n, A)$ for $n=2,3,\cdots,\infty$ extends by continuity to a derivation in $\mathrm{Der}(A_1, A)$, i.e. is the estimate (2.3.2) valid with $n_0=1$? A positive solution to this problem is suggested by Problem 2.1.3 . If the hypotesis that A is primitive is removed, the problem has a negative answer, see Theorem 2.4.20.
A positive answer to Problem 2.3.3 would also imply that all $\delta \in \mathrm{Der}(A_n, A)$ are closable as unbounded operators on A, at least when Δ comes from a Lie group action, by Theorem 2.3.6.

§2.3.2 Closability of derivations on A_1

We remarked in Definition 1.6.15, that in the case of derivations on C*-algebras the notion of dissipativeness can be formulated in other ways than in Definition 1.4.6. In the following Proposition we collect some of the characterizations which are called conservativity of δ.

Proposition 2.3.4

Let δ be a *-derivation on a C*-algebra A. The following conditions are equivalent

1. δ and $-\delta$ are dissipative (see Definition 1.4.6)
2. δ is conservative, i.e. for any $x = x^* \in D(\delta)$ there is a state ω on A such that $|\omega(x)| = \|x\|$ and $\omega(\delta(x)) = 0$
3. If $x = x^* \in D(\delta)$ and ω is a state on A such that $|\omega(x)| = \|x\|$, then $\omega(\delta(x)) = 0$
4. If $x = x^* \in D(\delta)$, there is a state ω on A such that $\omega(x) = \sup(\mathrm{Spec}(x))$ and $\omega(\delta(x)) = 0$
5. If $x = x^* \in D(\delta)$ and ω is a state on A such that $\omega(x) = \sup(\mathrm{Spec}(x))$, then $\omega(\delta(x)) = 0$
6. $\|(1 + \lambda\delta)(x)\| \geq \|x\|$ for any $x = x^* \in D(\delta)$ and any $\lambda \in \mathbb{R}$.

Proof

See [Bat 1].

Remark 2.3.5

Note that if $x = x^* \in A$, $x \neq 0$ and ω is a linear functional of norm 1 on A such that $\omega(x) = \sup(\mathrm{Spec}(x))$ (where $\mathrm{Spec}(x)$ is the Banach algebra spectrum of x), then ω is a state, see e.g. [BR 3], Lemma 1.

In proposition 2.3.4 we could have replaced the condition that δ is a derivation with the condition that A has an identity $\mathbb{1}$, and δ is a *-linear map with $\delta(\mathbb{1}) = 0$. If then in addition δ satisfies the range condition $(1 + \lambda\delta)(D(\delta)) = A$ for a positive and a negative λ, it follows from Theorem 1.5.2. that δ generates a one-parameter group of isometries with $e^{t\delta}(\mathbb{1}) = \mathbb{1}$, but this is then a one-parameter group of *-automorphisms and δ is a derivation, see e.g. [BR 1], Corollary 3.2.12. It would be interesting to see if the conditions of Proposition 2.3.4 together with $\delta(\mathbb{1}) = 0$ implies that δ is a derivation in general.

Note that if δ is not assumed to be a derivation, and if 5. is replaced by the conditions that δ is dispersive (i.e. if $x = x^* \in D(\delta)$ and $x_+ \neq 0$ (where x_+ is the positive part of x) and ω is a state with $\omega(x_+) = \|x_+\|$, then $\omega(\delta(x)) \geq 0$) and that $(1 + \lambda\delta)(D(\delta)) = A$ for some $\lambda > 0$, then $-\delta$ is the generator of a semigroup of positive contractions, [BR 3], Theorem 3. For a more systematic study of dispersiveness and related notions, see [BaR 1].

Note that even though Theorem 2.3.1 states that any derivation $\delta \in \mathrm{Der}(A_n, A)$ is continuous in the Frechet topology of A_n, it does not trivially follow that δ is closable as an unbounded operator on A. But

Theorem 2.3.6 (Batty)

Let Δ be a finite (or finite-dimensional) collection of conservative closed derivations on a C*-algebra A, and let A_1 be the corresponding *-algebra of once differentiable elements. Assume that A_1 is dense. If $\delta \in \mathrm{Der}(A_1, A)$, then δ is conservative, and in particular δ is closable.

Proof

This is a simplified version of the proof of Theorem 3 in [Bat 1]. If $x = x^* \in A_1$ we must show that $\omega(\delta(x)) = 0$ for all $x = x^* \in A_1$ and all states ω with $\omega(x) = \lambda = \inf(\mathrm{Spec}(x))$ (see condition 5 of Proposition 2.3.4). It follows from Condition 5 of Proposition 2.3.4 that we may assume that A has an identity, and then, subtracting $\lambda \mathbb{1}$ from x, we may assume that $\inf(\mathrm{Spec}(x)) = 0$. Then $x \geqslant 0$, and hence if $p \in A^{**}$ (= the enveloping von Neumann algebra of A) is the spectral projection of x corresponding to the spectral interval $\langle -\infty, 0]$, we have $pxp = 0$. But then $p\delta'(x)p = 0$ for all $\delta' \in \Delta$ by conservativity (If ϕ is a state of A with $\phi(p) > 0$ then $\omega(y) = \phi(p)^{-1}\phi(pyp)$ is a state of A with $\omega(x) = 0$, hence $\omega(\delta'(x)) = 0$ by conservativity and hence $p\delta'(x)p = 0$).

Let g_1, g_2, h be twice continuously differentiable real functions on $\mathbb{R}$ such that $g_1(0) = g_2(0) = 0$, $g_1(t)g_2(t) = t$ for $|t| \geqslant 1$, $h(t) = 0$ for $t \leqslant 1$, $h(t) = 1$ for $t \geqslant 2$, and put $f(t) = t - g_1(t)g_2(t)$. Then $f(nt)h(rt) = 0$ for $n \geqslant r \geqslant 0$. Define $x_n = n^{-1}f(nx)$, $n = 1,2,\ldots$. Then, as f is bounded, $\|x_n\| \xrightarrow[n\to\infty]{} 0$. Also, by Theorem 1.6.2, $x_n \in A_1$ and

$$\delta'(x_n) = (2\pi)^{-\frac{1}{2}}\int_{-\infty}^{\infty} ip\hat{f}(p)\left(\int^1 e^{intpx}\delta'(x)e^{in(1-t)px}dt\right)dp \quad \text{for all } \delta' \in \Delta,$$

and hence $\|\delta'(x_n)\| \leqslant C(\delta')$ for all n, where

$$C(\delta') = (2\pi)^{-\frac{1}{2}}\|\delta'(x)\| \int_{-\infty}^{\infty} |p\hat{f}(p)|dp \quad < +\infty .$$

Since $pe^{ikx} = e^{ikx}p = p$ for all $k \in \mathbb{R}$, it follows from the expression for δ' that $p\delta'(x_n)p = 0$ for all n.

Let y a weak*-limit point of the bounded sequence $\delta'(x_n)$ in $\mathfrak{A}^{**}$, so that $pyp = 0$. But as $\delta'(x_n h(rx)) = 0$ for $n \geqslant r$ and $\|x_n\| \to 0$ it follows from $\delta'(x_n h(rx)) = \delta'(x_n)h(rx) + x_n\delta'(h(rx))$ and closedness of δ' that $yh(rx) = 0$. But when $r \to \infty$, $h(rx) \to 1 - p$ σ-strongly and hence $y(1-p) = 0$. Thus $y = yp = py = pyp = 0$, i.e. any weak*-limit point of $\delta'(x_n)$ is 0. But then $\delta'(x_n)$ converges weak* to 0, and thus $\delta'(x_n)$ converges weakly to 0. As this is true for each $\delta' \in \Delta$ and Δ is finite, it follows from Hahn-Banach's separation theorem that there is a sequence x_n' in the convex hull of x_n such that $\|x_n'\| \to 0$, $\|\delta'(x_n')\| \to 0$ for all $\delta' \in \Delta$. But it follows from Theorem 2.3.1 that $\|\delta(y)\| \leqslant C(\|y\| + \sum_{\delta' \in \Delta}\|\delta'(y)\|)$ for $y \in A_1$ and a constant C, and hence $\|\delta(x_n')\| \to 0$. Now, let ω be a state such that $\omega(x) = 0 = \inf(\mathrm{Spec}\ x)$.

Since $g_1(0) = g_2(0) = 0$, it follows from spectral theory that $\omega(g_1(nx)^2) = \omega(g_2(nx)^2) = 0$, and as $x_n = x - n^{-1}g_1(nx)g_2(nx)$ we have $\delta(x - x_n) = n^{-1}\delta(g_1(nx))g_2(nx) + n^{-1}g_1(nx)\delta(g_2(nx))$. It follows from Schwarz's inequality that $\omega(\delta(x - x_n)) = 0$. This is still true in the convex hull of x_n: $\omega(\delta(x) - \delta(x_n')) = 0$, and therefore, in the limit $n \to \infty$, $\omega(\delta(x)) = 0$. Thus δ is conservative.

Problem 2.3.7

Does Theorem 2.3.6 remain valid if A_1 is replaced by A_n, $n = 2,3,\dots,\infty$, or by A_F in the case that Δ are the generators of the action of a compact Lie group? A partial answer to the last question is given in Theorem 2.3.8.

§2.3.3 Continuity on spectral-subspaces of derivations on A_F

We first consider compact groups, and next abelian groups.

Theorem 2.3.8 (Bratteli-Kishimoto)

Let G be a compact group, and let α be an action of G on a C^*-algebra A of operators acting non-degenerately on a Hilbert space H.

If $\delta \in \mathrm{Der}(A_F^\alpha, \mathcal{L}(H))$, then the restriction of δ to each of the spectral subspaces $A^\alpha(K)$ is bounded and σ-weakly continuous for each finite subset $K \subseteq \hat{G}$. Thus if the representation of A on H is G-covariant (in the sense of Definition 2.7.1), $M = A''$ and α also denotes the σ-weakly continuous extension of α to M, then δ has a unique σ-weakly continuous extension $\hat{\delta}$ from $A^\alpha(K)$ to $M^\alpha(K)$ such that

$$(2.3.6)\qquad \|\hat{\delta}|_{M^\alpha(K)}\| \leq \Big(\sum_{\gamma\in K} \dim(\gamma)^2\Big)\|\delta|_{A^\alpha(K)}\|$$

Proof

This is Theorem 2.1 in [BK 1].

If π is a finite-dimensional unitary representation of G on a d-dimensional space, define

$$A_R^\alpha(\pi) = \{x \in A \otimes M_d \,|\, \alpha_g(x) = (1 \otimes \pi(g)^*)\, x\, (1 \otimes \pi(g)) \text{ for all } g \in G\}$$

Then $A_R^\alpha(\pi)$ is a closed *-subalgebra of $\mathcal{L}(H) \otimes M_d$ which is contained in $A_F^\alpha \otimes M_d$, and thus

$$A_R^\alpha(\pi) \subseteq D(\delta \otimes 1)$$

By [Rin 1], see also Lemma 2.5.15, it follows that the restriction of the derivation $\delta \otimes 1$ to $A_R^\alpha(\pi)$ is bounded and σ-weakly continous. Now, put

$$\pi = 1 \oplus \sum_{\gamma \in K}^{\oplus} \gamma$$

Then all $\gamma \in K$ occurs as a sub-representation of $\bar{\pi} \otimes \pi$, namely as $1 \otimes \gamma$, and hence there exists an isometric injection of $A_1^\alpha(\gamma)$ into $A_R^\alpha(\pi)$ which is bicontinuous in the σ-weak topology, see Definition 2.2.12. Thus the map $\delta : A_1^\alpha(\gamma) \to L(H)^{\dim(\gamma)}$ is bounded and σ-weakly continuous for all $\gamma \in G$. Now any element $x \in A^\alpha(K)$ has the finite decomposition

$$x = \sum_{\gamma \in K} \sum_{i=1}^{d(\gamma)} P_{ii}(\gamma)(x)$$

by 2.2.35 and 2.2.40, and the maps

$$x \to [P_{ij}(\gamma)(x)]_{j=1}^{d(\gamma)} \in A_1^\alpha(\gamma)$$

are bounded and σ-weakly continous for $i = 1,\dots,d(\gamma)$, $\gamma \in K$. It follows that $\delta\big|_{A^\alpha(K)}$ is bounded and σ-weakly continuous.

Now, any $x \in M^\alpha(K)$ is the σ-weak limit of a net $x_\tau \in A^\alpha(K)$ such that $\|x_\tau\| \leq (\sum_{\gamma \in K} d(\gamma)^2)\|x\|$ by the following reasoning: By Kaplansky's density theorem, [BR 1], Theorem 2.4.16, there is a net $y_\tau \in A$ such that $\|y_\tau\| \leq \|x\|$ and $y_\tau \to x$ σ-weakly. For each τ, define

$$x_\tau = \sum_{\gamma \in K} \sum_{i=1}^{d(\gamma)} P_{ii}(\gamma)(y_\tau)$$

The maps $P_{ii}(\gamma) = d(\gamma) \int_G dg\, \overline{\gamma_{ii}(h)}\, \alpha_g$ are σ-weakly continous, and $\|P_{ii}\| \leq d(\gamma)$. It follows that $x_\tau \to x$ σ-weakly and $\|x_\tau\| \leq (\sum_{\gamma \in K} d(\gamma)^2)$

$\|x\|$. But as $\delta|_{A^\alpha(K)}$ is bounded, the net $\delta(x_\tau)$ is bounded and thus have a clustering point $y \in M^\alpha(K)$. As $\delta|_{A^\alpha(K)}$ is σ-weakly continous, we deduce that y is uniquely determined by x, and the definition $\hat{\delta}(x) = y$ defines a σ-weakly continuous extension of δ to $M^\alpha(K)$, and

$$\|\hat{\delta}(x)\| \leq \left(\sum_{\gamma\in K} d(\gamma)^2\right)\cdot\|\delta|_{A^\alpha(K)}\|\cdot\|x\|.$$

This ends the proof of Theorem 2.3.8.

Problem 2.3.9

Can the estimate 2.3.6 be improved to $\|\hat{\delta}|_{M^\alpha(K)}\| = \|\delta|_{A^\alpha(K)}\|$? The example on page 23-24 in [Sak 6] shows that this could be non-trivial, but it is true if $\delta(A_F^\alpha) \subseteq A$, by Theorem 2.3.12 and its proof. (See Remark 4.1)

Theorem 2.3.10 (Bratteli-Kishimoto)

Let G be a locally compact abelian group, and let α be an action of G on a C^*-algebra A.

If $\delta \in \mathrm{Der}(A_F^\alpha, A)$, and there exists a compact neighbourhood Ω of 0 in $\hat{G}$ such that the restriction of δ to $A^\alpha(\Omega)$ is bounded, then the restriction of δ to $A^\alpha(K)$ is bounded for each compact subset $K \subseteq \hat{G}$.

Remark 2.3.11

This is Theorem 3.1 in [BK 1], and we will follow the proof from there, which is similar to the proof of Theorem 2.3.1.

It is an open problem whether the assumption that $\delta|_{A^\alpha(\Omega)}$ is bounded is necessary for the conclusion. One case where this condition may be omitted is if A is simple and unital and $0 \in \hat{G}$ is isolated in the Connes spectrum $\Gamma(\alpha)$, see [BK 1, Theorem 3.2], and 2.6.1.

Proof of Theorem 2.3.10:

By a simple partition of unity argument on $\hat{G}$, it suffices to show that $\delta|_{A^\alpha(K)}$ is bounded for compact subsets $K \subseteq \hat{G}$ such that $K-K \subseteq \Omega$.

Define

$$L = \{a \in A \mid x \in A^{\alpha}(K) \to a\delta(x) \text{ is continuous}\}.$$

Then L is a left ideal in A, and L is closed by the uniform boundedness principle. If $y \in A^{\alpha}(K)$,

$$x \in A^{\alpha}(K) \to y^*\delta(x) = \delta(y^*x) - \delta(y^*)x$$

is continuous since $y^*x \in A^{\alpha}(K-K) \subseteq A^{\alpha}(\Omega)$. Thus $A^{\alpha}(K)^* \in L$. The closed linear span $\mathcal{D}$ of $A^{\alpha}(K)AA^{\alpha}(K)^*$ is an α-invariant hereditary C^*-subalgebra of A, and as

$$A^{\alpha}(K)AA^{\alpha}(K)^* \subseteq AA^{\alpha}(K)^* \cap A^{\alpha}(K)A \subseteq L \cap L^*$$

it follows that

$$\mathcal{D} \subseteq L \cap L^* \equiv B.$$

Here B is also a closed hereditary *-subalgebra of A.

Let

$$\mathcal{D}^{\perp} = \{a \in A \mid a\mathcal{D} = \mathcal{D}a = \{0\}\}$$

be the hereditary C^*-subalgebra of A orthogonal to $\mathcal{D}$. Since $\mathcal{D}^{\perp}$ is α-invariant, $\mathcal{D}^{\perp} \cap A_F^{\alpha}$ is dense in $\mathcal{D}^{\perp}$. For $a \in \mathcal{D}^{\perp} \cap A_F^{\alpha}$,

$$x \in A^{\alpha}(K) \to a\delta(x) = \delta(ax) - \delta(a)x = -\delta(a)x$$

is continuous. Thus $\mathcal{D}^{\perp} \subseteq L$ and then $\mathcal{D}^{\perp} \subseteq L \cap L^* = B$. As $\mathcal{D} \subseteq B$ it follows that B is an essential hereditary *-subalgebra in the sense that $B^{\perp} = \{0\}$.

Now we can prove that $x \in A^{\alpha}(K) \to \delta(x) \in A$ is closed. For suppose that x_n is a sequence in $A^{\alpha}(K)$ such that $x_n \to 0$, $\delta(x_n) \to z$. Then for $a \in B \subseteq L$, we have $a\delta(x_n) \to 0$. Thus $azz^* = 0$ and then $zz^*a^* = 0$, and $zz^* = 0$ since B is essential. It follows that $z = 0$, and $\delta|_{A^{\alpha}(K)}$ is continuous by the closed graph theorem. This ends the proof of Theorem 2.3.10.

We can furthermore prove that the derivation δ in Theorem 2.3.10 extends to the σ-weak closures of the spectral spaces $A^{\alpha}(K)$ in a G-covariant representation:

Theorem 2.3.12 (Ikunishi)

Let α be an action of a group G, which is compact or locally compact abelian, on a C*-algebra A. Let π be an G-covariant representation of A (see Definition 2.7.1), put $M = \pi(A)''$, and let $\hat{\alpha}$ denote the σ-weakly continuous representation of G in the automorphism group of M defined by $\hat{\alpha}_g(\pi(x)) = \pi(\alpha_g(x))$ for $x \in A$, $g \in G$.

Let $\delta \in \text{Der}(A_F^{\alpha}, A)$, and, if G is not compact, assume there exists a compact neighbourhood of Ω of 0 in $\hat{G}$ such that $\delta|_{A^{\alpha}(\Omega)}$ is bounded.

It follows that there exists a (unique) derivation $\hat{\delta} \in \text{Der}(M_F^{\hat{\alpha}}, M)$ such that

$$\hat{\delta}(\pi(x)) = \pi(\delta(x)) \tag{2.3.7}$$

for $x \in A_F^{\alpha}$,

$$\hat{\delta}|_{M^{\hat{\alpha}}(K)}$$

is bounded and σ-weakly continuous for each compact $K \subseteq \hat{G}$, and if G is not compact

$$\|\hat{\delta}|_{M^{\hat{\alpha}}(K)}\| \leq \inf_{\Omega} \|\delta|_{A^{\alpha}(K+\Omega)}\| \tag{2.3.8}$$

for each compact $K \in \hat{G}$, where Ω runs over all compact neighbourhoods of 0 in $\hat{G}$. If G is compact, then

$$\|\hat{\delta}|_{M^{\hat{\alpha}}(K)}\| \leq \|\delta|_{A^{\alpha}(K)}\| \tag{2.3.9}$$

for each finite subset $K \subseteq \hat{G}$.

Proof

This is Theorem 1 in [Iku 2].

First note that $\delta|_{A^{\alpha}(K)}$ is bounded for each compact subset $K \subseteq \hat{G}$, by Theorem 2.3.10. Let A^{**} be the bidual of A with the canonical von Neumann algebra structure, and let $\overline{A^{\alpha}(K)}$ be the closure of $A^\alpha(K)$ in A^{**} in the weak $*$-topology on A^{**}. Since $\delta|_{A^\alpha(K)}$ is bounded, its second adjoint δ^{**} is defined as a bounded weak$*$-continous operator from $\overline{A^\alpha(K)}$ into A^{**} for each compact $K \in \hat{G}$, with

$$\| \delta^{**}|_{\overline{A^\alpha(K)}} \| = \| \delta|_{A^\alpha(K)} \| .$$

As $K_1 \subseteq K_2 \Rightarrow \overline{A^\alpha(K_1)} \subseteq \overline{A^\alpha(K_2)}$, δ^{**} thus define an operator on the space $C = \bigcup_K \overline{A^\alpha(K)} \subseteq A^{**}$ (but we do not claim that δ^{**}, so defined, is a restriction of the bi-adjoint of δ).

We now argue that C is a *-algebra and δ^{**} is a derivation. From now, G will generally be assumed to be locally compact abelian, and the simpler argument when G is compact will be described parenthetically.

Since $A^\alpha(K_1)A^\alpha(K_2) \subseteq A^\alpha(K_1+K_2)$ for any pair K_1, K_2 of compacts in $\hat{G}$, 2.2.47, and

$$\overline{A^\alpha(K_1+K_2)} = \left\{ x \in A^{**} \mid \eta(x) = 0 \ \text{ for all } \ \eta \in A^* \cap A^\alpha(K_1+K_2)^\perp \right\}$$

it follows by limiting that

$$\text{(2.3.10)} \quad \overline{A^\alpha(K_1)}\;\overline{A^\alpha(K_2)} \subseteq \overline{A^\alpha(K_1+K_2)}$$

Also $\overline{A^\alpha(K)}^* = \overline{A^\alpha(K)^*} = \overline{A^\alpha(-K)}$, thus C is a *-algebra. Then δ^{**} is a *-derivation by limiting. (If G is compact, use $A^\alpha(\gamma_1)A^\alpha(\gamma_2) \subseteq A^\alpha$ (irreducible components of $\gamma_1 \otimes \gamma_2$).)

If B is the norm closure of C, then B is a C^*-subalgebra of A^{**}, invariant under the second adjoint α^{**} of α since the subspaces $\overline{A^\alpha(K)}$ are so. As the mapping $g \in G \to \alpha_g|_{A^\alpha(K)}$ is norm continous, so is $g \in G \to \alpha_g^{**}|_{A^\alpha(K)^{**}} = \alpha_g^{**}|_{\overline{A^\alpha(K)}}$, where we have done the identification $A^\alpha(K)^{**} = \overline{A^\alpha(K)}$ by Hahn-Banach. Thus the restriction $\alpha^{**}|_B$ is a strongly continuous action of G on B which we from now will denote by α.

We shall show that

$$\text{(2.3.11)} \quad \overline{A^\alpha(K)} \subseteq B^\alpha(K) = \bigcap_\Omega \overline{A^\alpha(K+\Omega)}$$

for all compact $K \in \hat{G}$, where Ω runs over all compact neighbourhoods of 0 in $\hat{G}$. The first inclusion is trivial from the definition of $B^{\alpha}(K)$ as the hull of the set of α_f where $f \in L^1(G)$ and $\hat{f} = 0$ on K, Definition 2.2.5. For the last equality, choose a $f \in L^1(G)$ such that $\operatorname{supp} \hat{f} \subseteq K+\Omega$ and $\hat{f} = 1$ on K, then

$$B^{\alpha}(K) \subseteq \alpha_f(B) = \alpha_f^{**}(B) \subseteq \overline{\alpha_f(A)}$$
$$\subseteq \overline{A^{\alpha}(K+\Omega)} \subseteq B^{\alpha}(K+\Omega)$$

where α_f^{**} is the bidual of $\alpha_f\big|_A$. By Proposition 2.2.8, $B^{\alpha}(K) = \bigcap_{\Omega} B^{\alpha}(K+\Omega)$ and 2.3.11 follows. (If G is compact, we have simply $B^{\alpha}(K) = \overline{A^{\alpha}(K)}$.)

In particular, 2.3.11 implies that $B^{\alpha}(K)$ is closed in the σ-weak topology on A^{**}.

Let $\tilde{\pi}$ be the canonical extension of π to A^{**}, and put

(2.3.12) $\quad I = \ker \tilde{\pi} \cap B$

We now argue that the ideal I in B has an identity e, which is then a central projection in B. Let e_τ be an approximate identity for I. Now if f is a function in $L^1(G)$ such that $f \geq 0$, $\int_G dg f(g) = 1$ and $\operatorname{supp} \hat{f} = K$ is compact, then $\tau \to \alpha_f(e_\tau)$ is still an approximate identity, and $\alpha_f(e_\tau) \subseteq B^{\alpha}(K)$ for all τ. Now $\alpha_f(e_\tau)$ converges strongly to a projection $e \in A^{**}$, but as $B^{\alpha}(K)$ is a σ-weakly closed subspace of A^{**}, it follows that $e \in B^{\alpha}(K)$. But as $\tilde{\pi}(\alpha_f(e_\tau)) = 0$ for all τ, it follows that $\tilde{\pi}(e) = 0$, i.e. $e \in I$. As $\alpha_f(e_\tau)$ is an approximate identity for I, it follows that e is an identity for I, hence $e \in$ centre (B) and

(2.7.84) $\quad I = eB = Be.$

(Note that as $\tilde{\pi}$ is a α^{**}-covariant representation of B, we actually have $e \in B^{\alpha}$.) It follow for any compact $K \subset \hat{G}$ that $B^{\alpha}(K)(1-e)$ and $\tilde{\pi}(B^{\alpha}(K))$ are isometrically isomorphic, but then they are also σ-weakly homeomorphic since $\tilde{\pi}$ is a normal representation of A^{**}, and both spaces are σ-weakly closed with compact unit balls. (Use Krein-Smulyan's theorem.) Further, from the previous regularization argument:

$$\begin{aligned}M^{\hat{\alpha}}(K) &= \bigcap_{\Omega} \overline{\pi(A^{\alpha}(K+\Omega))} \\ &= \tilde{\pi}(\bigcap_{\Omega} \overline{A^{\alpha}(K+\Omega)}) \\ &= \tilde{\pi}(B^{\alpha}(K))\end{aligned}$$

where Ω runs over all compact neighbourhoods of 0 in $\hat{G}$. Thus $B^{\alpha}(K)(1-e)$ is isometrically isomorphic and σ-weakly homeomorphic to $M^{\hat{\alpha}}(K)$ under $\tilde{\pi}$.

Returning to δ^{**}, since $e \in$ centre (B) and B is dense in A^{**}, we have $e \in$ centre (A^{**}) and as δ^{**} is a derivation, it follows that

$$\begin{aligned}\delta^{**}(e)x &= \delta^{**}(ex) - e\delta^{**}(x) \\ &= \delta^{**}(xe) - \delta^{**}(x)e \\ &= x\delta^{**}(e)\end{aligned}$$

for all $x \in C$, and then for all $x \in A^{**}$. It follows that $\delta^{**}(e) \in$ centre (A), and since e is a projection, it follows as in the proof of Lemma 1.6.9 that

$$\delta^{**}(e) = 0$$

But then

$$\delta^{**}(x(1-e)) = \delta^{**}(x)(1-e)$$

for all $x \in G$.

Next, let $p \in A^{**}$ be the central projection corresponding to the representation π, i.e. p is the range-projection of the ideal $\ker(\tilde{\pi})$ in A^{**}. As $\tilde{\pi}(e) = 0$, we have

$$p \geqslant e$$

(Note that we may have $p \neq e$, if for example $G = \mathbb{T}^2$ and α is an ergodic action of G on a simple C^*-algebra A, see Remark 2.9.2, then $B = A$ and $e = 0$ for any representation π, while p depends on π.) If A is embedded canonically into A^{**}, the representation $\tilde{\pi}$ is quasi-equivalent with the representation

$$x \in A^{**} \to x(1-p)$$

and we may identify $\tilde{\pi}$ with the latter representation. We have already remarked that the map

$$x(1-e) \in B^{\alpha}(K)(1-e) \to x(1-p) \in M^{\hat{\alpha}}(K)$$

is an σ-weakly bi-continous isometric isomorphism. Thus, if we define a derivation $\hat{\delta}$ on $M_F^{\hat{\alpha}} \cong B_F^{\alpha}(1-e)$ by

$$\hat{\delta}(x(1-p)) = (1-p)\delta^{**}(x)$$

for $x \in G$, then $\hat{\delta}$ is well defined by the following argument: If $x \in B^{\alpha}(K)$ and $x(1-p) = 0$, then $x(1-e) = 0$, thus $\delta^{**}(x)(1-e) = \delta^{**}(x(1-e)) = 0$ and hence $\delta^{**}(x)(1-p) = 0$. Also $\hat{\delta}$ is a derivation since p is a central projection. The restriction $\hat{\delta}|_{M^{\hat{\alpha}}(K)}$ is σ-weakly continous and it is bounded by the relations

$$\hat{\delta}(x(1-p)) = (1-p)\delta^{**}(x(1-e))$$

and

$$\|x(1-p)\| = \|x(1-e)\|$$

for $x \in B^{\alpha}(K)$, and thus

$$\begin{aligned} \|\hat{\delta}|_{M^{\hat{\alpha}}(K)}\| &\leq \|\delta^{**}|_{B^{\alpha}(K)(1-e)}\| \\ &\leq \inf_{\Omega} \|\delta^{**}|_{\overline{A^{\alpha}(K+\Omega)}}\| \\ &\leq \inf_{\Omega} \|\delta|_{A^{\alpha}(K+\Omega)}\| . \end{aligned}$$

Hence $\hat{\delta}$ is the sought after derivation in Theorem 2.3.12.

§2.4 Abelian C*-algebras

It is well known from Gelfand theory that abelian C*-algebras A has the form $A = C_0(\Omega)$ = the continuous functions on Ω vanishing at infinity, where Ω is the spectrum of A, i.e. Ω is the set of multiplicative functionals on A equipped with the weak *-topology. The space Ω is a locally compact Hausdorff space, [BR 1], Chapter 2.3.5. Conversely, if Ω is a locally compact Hausdorff space, $A = C_0(\Omega)$ is an abelian C*-algebra with spectrum Ω.

There is a canonical isomorphism between the group Aut (A) of *-automorphisms of A and the group Homeo (Ω) of homeomorphisms of Ω, given by

(2.4.1) $$(\alpha f)(\omega) = f(T^{-1}\omega)$$

for all $f \in C_0(\Omega)$, $\omega \in \Omega$. Here $\alpha \in$ Aut (A), $T \in$ Homeo (Ω), and α and T determines each other uniquely by the relation 2.4.1. This correspondence has the following continuity property.

Lemma 2.4.1

Let $A = C_0(\Omega)$ be an abelian C*-algebra, G a topological group, $g \in G \to \sigma_g \in$ Aut (A) a representation of G as *-automorphisms of A, and $g \to S_g \in$ Homeo (Ω) the corresponding representation of G as homeomorphisms of Ω. The following conditions are equivalent:

1. The map $g \in G \to \sigma_g$ is strongly continuous, i.e. $\|\sigma_g(f) - f\| \xrightarrow[g\to e]{} 0$ for all $f \in A$.
2. The map $(g,\omega) \in G \times \Omega \to S_g\omega \in \Omega$ is jointly continuous.

Proof

$1 \Rightarrow 2$: Assume 1, and let $(g_\alpha, \omega_\alpha)$ be a net in $G \times \Omega$ converging to (g,ω). If $f \in C_0(\Omega)$, we have

$$|f(S_{g_\alpha}\omega_\alpha) - f(S_g\omega)|$$

$$\leq \|\sigma_{g_\alpha^{-1}}(f) - \sigma_{g^{-1}}(f)\| + \|(\sigma_{g^{-1}}f)(\omega_\alpha) - (\sigma_{g^{-1}}f)(\omega)\| \xrightarrow[\alpha\to\infty]{} 0 + 0 = 0,$$

and as Ω is locally compact, it follows that $\lim_{\alpha} S_{g_\alpha}\omega_\alpha = S_g\omega$.

$2 \Rightarrow 1$: Assume that 1 is false, i.e. that there exists an $f \in C_0(\Omega)$ such that $\overline{\lim_{g\to e}} \|\sigma_g(f) - f\| = 2\varepsilon > 0$. Then there exists a net $(g_\alpha, \omega_\alpha) \in G \times \Omega$ such that $g_\alpha \to e$ and

$$(2.4.2) \quad |f(S_{g_\alpha}\omega_\alpha) - f(\omega_\alpha)| \geq \varepsilon$$

for all α. But the set $K = \{\omega \in \Omega | \; |f(\omega)| \geq \varepsilon/2\}$ is compact, and, replacing (g_α,ω_α) by $(g_\alpha^{-1}, S_{g_\alpha}\omega_\alpha)$ if necessary, we may assume that $\omega_\alpha \in K$ for all α. By compactness, there exists a subnet of (g_α,ω_α) and an $\omega \in K$ such that $g_\alpha \to e$, $\omega_\alpha \to \omega$ if $\alpha \to \infty$ in the subnet. But because of 2.4.2, we cannot have $\lim_{\alpha\to\infty} S_{g_\alpha}\omega_\alpha = \omega$ over the subnet, and 2 does not hold.

The previous lemma makes the following definition relevant in a C*-setting.

Definition 2.4.2

Let Ω be a locally compact Hausdorff space, and G a topological group. A mapping $g \in G \to S_g \in \text{Homeo}(\Omega)$ is called a G-flow if

1. S is a representation, i.e. $S_e = 1 \quad S_gS_h = S_{gh}$, $g,h \in G$

2. The map $(g,\omega) \in G \times \Omega \to S_g\omega \in \Omega$ is jointly continuous.

A $\mathbb{R}$ - flow will simply be called a flow, and the generator of the corresponding one parameter group σ of automorphisms;

$$(2.4.3) \quad (\sigma_t f)(\omega) = f(S_t\omega)$$

will be called the generator of the flow.

The generator of a flow can also be characterized as a derivative in the primitive sense. This is analogous to the equivalence of weak and strong generators of a strongly continuous semigroup of bounded linear maps, see [BR 1], Corollary 3.1.8.

Lemma 2.4.3

Let S be a flow on Ω with generator δ, and σ the corresponding one-parameter group of $*$-automorphisms of $C_0(\Omega)$. Let $f \in C_0(\Omega)$.

The following conditions are equivalent

1. There exists a $g \in C_0(\Omega)$ such that

(2.4.4) $\lim_{t\to 0} \|(\sigma_t(f) - f)/t - g\| = 0$

2. There exists a $g \in C_0(\Omega)$ such that

(2.4.5) $g(\omega) = \lim_{t\to 0} (f(S_t\omega) - f(\omega))/t$ for each $\omega \in \Omega$

If 1 or 2 is fullfilled, then $f \in D(\delta)$ and $\delta(f) = g$.

Proof

$1 \Rightarrow 2$ and the last statement are trivial. Assume 2. Then by the fundamental theorem of calculus $f(S_t\omega) - f(\omega) = \int_0^t ds\, g(S_s\omega)$ and hence $|(f(S_t\omega) - f(\omega))/t - g(\omega)| = |\frac{1}{t}\int_0^t ds (g(S_s\omega) - g(\omega))|$ $\leq \sup_{|s|\leq|t|} \| \sigma_s(g) - g \|$. But the last number is independent of ω and tends to zero as $t \to 0$ by strong continuity. Thus $f \in D(\delta)$ and $\delta(f) = g$.

§2.4.1 Classification of derivations

We will consider the following situation $A = C_0(\Omega)$ is an abelian C*-algebra with spectrum Ω; $\Delta = \{\delta_1, \ldots, \delta_d\}$ is a finite set of closed derivations on A; A_n = the algebra of n times differentiable elements in A, $n = 1,2, \ldots, \infty$, see 2.3.1.

We will show under general circumstances that if $\delta \in \mathrm{Der}(A_n, A)$, then δ has the form $\delta = \sum_{i=1}^{d} \ell_i \delta_i$ where ℓ_i are functions on Ω. To this end we need several lemmas.

Lemma 2.4.4

Assume that A_n is dense. If $\omega_0 \in \Omega$ and V is a neighborhood of ω_0, there exists a $g \in A_n$ such that $g = 1$ in a neighbourhood of ω_0 and $\operatorname{supp} g \subseteq V$.

Here

(2.4.6) $\operatorname{supp} g$ = the closed support of $g = \overline{\{\omega \in \Omega | g(\omega) \neq 0\}}$.

Proof

Since A_n is dense, there exists a $f \in A_n$ such that $f(\omega_0) = 1$, and $f(\omega) < \frac{1}{4}$ for all $\omega \in \Omega \setminus V$. Let h be a function in $C^{\infty}_{\mathbb{R}}(\mathbb{R})$ such that $h(t) = 0$ for $|t| < \frac{1}{2}$, $h(t) = 1$ for $|t-1| < \frac{1}{4}$. Then $g = h \circ f$ is in A_n by lemma 2.3.2 and g has the desired properties.

The next lemma states that derivations has an important locality property.

Lemma 2.4.5

Assume that A_n is a dense and let $\delta \in \operatorname{Der}(A_n, A)$. Then $\operatorname{supp} \delta(f) \subseteq \operatorname{supp} f$ for all $f \in A_n$.

Proof

Suppose that $f \in A_n$ and that $f = 0$ in a neighborhood V of ω_0 where $\omega_0 \in \Omega$. Then, by lemma 2.4.4 there exists a $g \in A_n$ such that $g(\omega_0) = 1$ and $\operatorname{supp} f \subseteq V$. As $f \cdot g = 0$, we have
$0 = \delta(f \cdot g)(\omega_0) = (\delta f)(\omega_0) g(\omega_0) + (\delta g)(\omega_0) f(\omega_0) = (\delta f)(\omega_0)$
i.e. $\delta f(\omega_0) = 0$. Thus $\operatorname{supp}(\delta(f)) \subseteq \operatorname{supp}(f)$.

Lemma 2.4.6 (Batty)

Assume A_n is dense. Fix $\omega_0 \in \Omega$. Let f be an element of A_{2n} which satisfies

(2.4.7) $(\delta_{i_1}, \dots \delta_{i_j} f)(\omega_0) = 0$ for $0 \leq j \leq n$, $i_k \in \{1 \dots d\}$.

Then there exists a sequence $\{f_m\}_{m=0}^{+\infty} \subset A_n$ such that each f_m is zero in a neighborhood of ω_0 and such that $\|f - f_m\|_n \to 0$ $(m \to +\infty)$, where $\|\ \|_n$ is the Banach algebra norm 2.2.7 on A_n if $n < \infty$, and $\|\ \|_n$ is any Fréchet norm on A_∞ if $n = \infty$.

Remark 2.4.7

This is a slightly extended version of Proposition 5.2 in [Bat 4]. It is unknown whether it suffices to assume $f \in A_n$ except in the cases $n = 1$, see Proposition 5.1 in [Bat 4], and $n = \infty$.

Proof

The case that $n = \infty$ can be treated by trivial modifications of the following argument, so let n be finite. We may assume that

(2.4.8) $\|\delta_{i_1} \cdots \delta_{i_j} f\| \leq 1$, $\quad 0 \leq j \leq 2n$.

Let g be an element of $C^n(\mathbb{R})$ such that $0 \leq g(t) \leq 1$ $(t \in \mathbb{R})$, $g(t) = 1$ in a neighborhood of 0, $g(t) = 0$ for $|t| \geq 1$. For $0 < \varepsilon < 1$, put

(2.4.9) $$\varepsilon_j = \begin{cases} \varepsilon^{n+1-j} & \text{for } 0 \leq j \leq n, \\ 1 & \text{for } n < j \end{cases}$$

Define

(2.4.10) $$h_{\varepsilon,i_1,\ldots,i_j} = g \circ (\varepsilon_j^{-1} \delta_{i_1} \cdots \delta_{i_j} f) \quad (0 \leq j \leq n),$$

(2.4.11) $$h_\varepsilon = \prod_{j=0}^{n} \prod_{i_1=1}^{d} \cdots \prod_{i_j=1}^{d} h_{\varepsilon,i_1,\ldots,i_j}, \text{ and}$$

(2.4.12) $$f_\varepsilon = f(1 - h_\varepsilon).$$

Then the properties of g and f imply that $h_{\varepsilon,i_1,\ldots,i_j}$, h_ε and f_ε belong to A_n, and $f_\varepsilon = 0$ in a neighborhood of ω_0 by 2.4.7 and 2.4.10. We will prove the lemma by showing

(2.4.13) $$\lim_{\varepsilon \to 0} \|f - f_\varepsilon\|_n = 0,$$

that is,

(2.4.14) $\lim_{\varepsilon\to 0} \|fh_\varepsilon\|_n = 0$.

By Example 1.6.1, applied to 2.4.10:

$$\delta_{k_r}(h_{\varepsilon,i_1,\ldots,i_j}) = \varepsilon_j^{-1}\, g\circ(\varepsilon_j^{-1}\delta_{i_1} \cdots \delta_{i_j} f)\delta_{k_r}\delta_{i} \cdots \delta_{i_j} f ,$$

$$\delta_{k_{r-1}}\delta_{k_r}(h_{\varepsilon,i_1,\ldots,i_j})$$

$$= \varepsilon_j^{-2} g''(\varepsilon_j^{-1}\delta_{i_1} \cdots \delta_{i_j} f)\delta_{k_{r-1}}\delta_{i_1} \cdots \delta_{i_j} f \cdot \delta_{k_r}\delta_{i_1} \cdots \delta_{i_j} f$$

$$+ \varepsilon_j^{-1} g'\circ(\varepsilon_j^{-1}\delta_{i_1} \cdots \delta_{i_j} f)\delta_{k_{r-1}}\delta_{k_r}\delta_{i_1} \cdots \delta_{i_j} f .$$

Proceeding recursively, we have

(2.4.15) $\delta_{k_1} \cdots \delta_{k_r}(h_{\varepsilon,i_1,\ldots,i_j})$

= a sum of N_r terms of the form $F_{i_1,\ldots,i_j,\, k_1,\ldots,k_r,\, S}$,

where $F_{i_1,\ldots,i_j,\, k_1,\ldots,k_r,\, S}$

$$= \varepsilon_j^{-|S|} g^{(|S|)} \circ (\varepsilon_j^{-1}\delta_{i_1} \cdots \delta_{i_j} f) \prod_{K\in S} \delta_K \delta_{i_1} \cdots \delta_{i_j} f ,$$

S is a partition of the ordered set $\{k_1 \ldots k_r\}$ into $|S|$ subsets, the elements in K is ordered as in $k_1, \ldots, k_r$, $\delta_k = \delta_{m_1} \cdots \delta_{m_s}$ if $K = \{m_1 \ldots m_s\}$, and N_r is a number depending only on r. We then have

(2.4.16) $N_r \le N_{r+1}$, $r = 0, 1, 2, \ldots$.

Let us estimate $F_{i_1,\ldots,i_j,\, k_1,\ldots,k_r,\, S}$. Define

(2.4.17) $V_\varepsilon \equiv \{\omega\in\Omega; |\delta_{i_1} \cdots \delta_{i_j} f(\omega)| \le \varepsilon_j \text{ for } 0\le j\le n,\ i_k\in\{1\ldots d\}\}$

$$= \{\omega\in\Omega; |\delta_{i_1} \cdots \delta_{i_j} f(\omega)| \le \varepsilon_j \text{ for } 0\le j\le 2n,\ i_k\in\{1\ldots d\}\},$$

where the last equality follows from 2.4.8 and 2.4.9. Also, define

(2.4.18) $M = \sup\{\|g^{(j)}\| ; j = 0, \ldots n\}$.

Using 2.4.9, we have $\varepsilon_j^{-1}\varepsilon_{j+k} \le \varepsilon^{-k}$ for $j, k \ge 0$. Thus, for $\omega\in V_\varepsilon$:

$$|F_{i_1,\dots,i_j,\ k_1,\dots,k_r,\ S}(\omega)| \leq \varepsilon_j^{-|S|} M \prod_{K \in S} \varepsilon_{j+|K|}$$

$$= M \prod_{K \in S} (\varepsilon_j^{-1}\ \varepsilon_{j+|K|})$$

$$= M\ \varepsilon^{-r} \leq (M\varepsilon^{-1})^r \quad ,$$

where $|K|$ is the number of elements of K, and we used $M \geqslant 1$.

Thus, from 2.4.15,

(2.4.19) $$|\delta_{k_1} \dots \delta_{k_r}(h_{\varepsilon,i_1,\dots,i_j})(\omega)| \leq N_r (M\varepsilon^{-1})^r \quad \text{for} \quad \omega \in V_\varepsilon \ .$$

It follows from 2.4.11 that $\delta_{k_1}\delta_{k_2} \dots \delta_{k_r}(h_\varepsilon)$ is the sum of N_r' terms, each of which is a product of functions $h_{\varepsilon,j_1,\dots,j_\ell}$ together with the derivatives of these, and the sum of the order of the derivatives of each term is r. Here N_r' is a number only depending on r and d. As $\|h_{\varepsilon,\ j_1,\dots,j_\ell}\| \leqslant 1$, we get the following estimate from 2.4.19:

(2.4.20) $$|\delta_{k_1} \dots \delta_{k_r}(h_\varepsilon)(\omega)| \leqslant N_r' N_r (M\varepsilon^{-1})^r \quad \text{for} \quad \omega \in V_\varepsilon \ .$$

Finally, $\delta_{k_1} \dots \delta_{k_r}(fh_\varepsilon)$ is a sum of 2^r terms, and there are $\binom{r}{\ell}$ terms which are the product of an ℓ'th order derivative of f with an $(r-\ell)$'th order derivative of h_ε . It follows from 2.4.8, 2.4.9, 2.4.16, 2.4.20, and $N_\ell' < N_{\ell+1}'$, that if $\omega \in V_\varepsilon$, then:

$$|\delta_{k_1} \dots \delta_{k_r}(fh_\varepsilon)| \leq \sum_{\ell=0}^{r} \binom{r}{\ell} \varepsilon_\ell\ N'_{r-\ell} N_{r-\ell} (M\varepsilon^{-1})^{r-\ell}$$

$$\leq N_r' N_r M^r \sum_{\ell=0}^{r} \binom{r}{\ell} \varepsilon^{n+1-\ell} \varepsilon^{-r+\ell}$$

$$= N_r' N_r (2M)^r \varepsilon^{n+1-r} \quad .$$

Hence

(2.4.21) $$\sup_{\omega \in V_\varepsilon} |\delta_{k_1} \dots \delta_{k_r}(fh_\varepsilon)(\omega)| \leq N_r' N_r (2M)^r\ \varepsilon^{n+1-r} \quad .$$

If $\omega \notin V_\varepsilon$ it follows from the definition of g, 2.4.10 and 2.4.17 that there exists $0 \leqslant j \leqslant n$ and $i_k \in \{1, \dots, d\}$ such that

$h_{\varepsilon, i_1, \ldots, i_j}(\omega) = 0$, thus $h_\varepsilon(\omega) = 0$ and $fh_\varepsilon(\omega) = 0$. But as V_ε is closed by 2.4.17, the complement of V_ε is open, and it follows from Lemma 2.4.5 that $\delta_{k_1} \ldots \delta_{k_r}(fh_\varepsilon)(\omega) = 0$ for $\omega \in \Omega \setminus V_\varepsilon$ and $r = 0, 1, \ldots, n$. Thus, from 2.4.21

$$\|\delta_{k_1} \ldots \delta_{k_r}(fh_\varepsilon)\| \leq N_r{}'N_r(2M)^r\varepsilon^{n+1-r} \quad \text{for } r = 0, 1, \ldots, n,$$

$k_m \in \{1, \ldots, d\}$. Thus $\|\delta_{k_1} \ldots \delta_{k_r}(fh_\varepsilon)\| \to 0$ as $\varepsilon \to 0$, and this establishes 2.4.14 and the lemma.

Problem 2.4.8

Is the conclusion of Lemma 2.4.6 valid if one assumes $f \in A_n$ instead of $f \in A_{2n}$?

The next lemma is an extension of Proposition 2.2.1.

Lemma 2.4.9

Let A be a Banach space, G a Lie group, σ an action of G on A as a strongly continuous group of isometries, A_n the space of n times differentiable elements for $n = 0, 1, \ldots, \infty$. If $0 \leqslant n \leqslant m \leqslant \infty$, then A_m is dense in A_n in the Fréchet topology of A_n .

Proof

Let $d\sigma$ be the Lie algebra representation of the Lie algebra $\mathfrak{g}$ of G corresponding to the group representation σ, i.e.

(2.4.22) $d\sigma(X)$ = infinitesmal generator of the one-parameter group $t \in \mathbb{R} \to \sigma(\exp\{tX\})$ for each $X \in \mathfrak{g}$. Then

(2.4.23) $$A_n = \bigcap\{D(d\sigma(X_1) \ldots d\sigma(X_m)) \,|\, m < n+1, X_k \in \mathfrak{g}\}$$

and the topology on A_n is determined by the seminorms.

(2.4.24) $$x \in A_n \to \|d\sigma(X_1) \ldots d\sigma(X_m)\, x\| \quad \text{for } m < n+1, X_k \in \mathfrak{g}.$$

Now the adjoint representation of G on the finite dimensional space $\mathfrak{g}$, determined by

(2.4.25) $\exp\{Ad(g)(X)\} = g \exp\{X\} g^{-1}$
for $g \in G$, $X \in \mathfrak{g}$ is continuous. But

(2.4.26) $d\sigma(Ad(g)(X)) = \sigma_g d\sigma(X)\sigma_g^{-1}$.

It follows that if $x \in A_1$ then

(2.4.27) $\|\sigma_g(d\sigma(X)x) - d\sigma(X)(\sigma_g x)\|$

$= \|d\sigma(X)x - \sigma_g^{-1} d\sigma(X)\sigma_g x\|$

$= \|d\sigma(X - Ad(g^{-1})(X))x\| \to 0$ as $g \to e$

for all $X \in \mathfrak{g}$, where the last relation follows from finite-dimensionality of $\mathfrak{g}$. Similarly since the set of polynomials of degree less than a finite k in elements of $\mathfrak{g}$ is a finite-dimensional subspace of the enveloping associative algebra of $\mathfrak{g}$, one deduces that

(2.4.28) $\|\sigma_g d(X_1) \ldots d\sigma(X_k)x - d\sigma(X_1) \ldots d\sigma(X_k)\sigma_g x\| \to 0$

as $g \to e$ for all $x \in A_k$.

Now let $h_m \in C_{00}^\infty(G)$ be a sequence of positive functions of integral 1 such that for any neighborhood V around e, supp $h_m \subseteq V$ for sufficiently large m. If $f \in A_n$, define $f_m = \int_G dg\, h_m(g)\sigma_g(f)$. Then clearly $f_m \in A_\infty$, and if $k < n + 1$ and $X_1, \ldots, X_k \in \mathfrak{g}$, then

(2.4.29) $\|d\sigma(X_1) \ldots d\sigma(X_k)(f_m - f)\|$

$= \|\int_G dgh_m(g)d\sigma(X_1) \ldots d\sigma(X_k)(\sigma_g(f) - f)\|$

$< \int_G dgh_m(g)\|(d\sigma(X_1) \ldots d\sigma(X_k)\sigma_g f - \sigma_g d\sigma(X_1) \ldots d\sigma(X_k)f)\|$

$+ \int_G dgh_m(g)\|(\sigma_g - 1)d\sigma(X_1) \ldots d\sigma(X_k)f\| \to 0$

as $g \to 0$ by 2.4.28 and strong continuity of σ. This establishes the lemma.

The next lemma tells that smooth functions on compact subsets of G-orbits in Ω extends to smooth functions on Ω.

Lemma 2.4.10

Let G be a Lie group, Ω a locally compact Hausdorff space, S a G-flow on Ω, and for a fixed $\omega \in \Omega$ define the stabilizer group at ω as the closed subgroup

(2.4.30) $G_\omega = \{g \in G \mid S_g\omega = \omega\}$

of G. Then G/G_ω is a C^∞-manifold, and if $h \in C_{00}^\infty(G/G_\omega)$ and K is a compact subset of G/G_ω, then there exists an $f \in C_{00}^\infty(\Omega)$ such that

(2.4.31) $f(S_g\omega) = h(g/G_\omega)$

for $g \in G$ such that $g/G_\omega \in K$. Here $C_{00}^\infty(\Omega)$ denotes the continuous functions with compact support in Ω which are infinitely differentiable with respect to the action of G, and g/G_ω denotes the image of $g \in G$ in G/G_ω under the quotient map.

Proof

G_ω is a Lie group and G/G_ω is a manifold by general Lie group theory, and G acts on the Banach space $C_0(G/G_\omega)$ by left translation. As $h \in C_0(G/G_\omega)$, it follows from theorem 2.2.3 that if V is an arbitrary compact neighborhood of e in G, there exists finite sequences $\psi_i \in C_{00}^\infty(G)$, $h_i \in C_{00}(G/G_\omega)$ with $\operatorname{supp}\psi_i \in V$ such that

(2.4.32) $h(g/G_\omega) = \sum_{i=1}^{n} \int_G dg'\psi_i(g')h_i(g'^{-1}g/G_\omega)$

for all $g/G_\omega \in K$. (Actually Theorem 2.2.3 says that we can obtain this relation for all g/G_ω if we only assume $h_i \in C_0(G/G_\omega)$, but by modifying h_i outside the compact set $\overline{VK}$, the expression 2.4.32 is still valid for $g/G_\omega \in K$.) Let K' be a compact subset of G/G_ω such that $\operatorname{supp} h_i \subseteq K'$ for $i = 1, \ldots, n$ and $VK \subseteq K'$. The map $g/G_\omega \in K' \to S_g\omega$ is a one-one continuous map from a compact Hausdorff space into a locally compact Hausdorff space, thus it is a homeomorphism on the range. We may thus view h_i as continuous function on the compact range of this map, by transporting h_i by the map. By Tietze's extension theorem, these functions extends to functions in $C_{00}(\Omega)$ which we will also denote by h_i. Define

$$(2.4.33)\quad f(\omega') = \sum_{i=1}^{n} \int_G dg\psi_i(g)h_i(S_{g^{-1}}\omega')$$

for all $\omega' \in \Omega$, where h_i are the extended functions. Then $f \in C_0^\infty(\Omega)$ since $\psi_i \in C_{00}^\infty(\Omega)$ (Use Lemma 2.4.3 or note that $f = \sum_{i=1}^{n} \int_G dg\, \psi_i(g)\sigma_g(h_i)$ where σ is the automorphism group associated to S), and f has compact support in $\overline{\bigcup_{i=1}^{n} S_{\mathrm{supp}\psi_i}(\mathrm{supp}\ h_i)}$ (the latter sets are compact because of the joint continuity of $(g,\omega) \to S_g\omega)$. One easily verifies from 2.4.32 and 2.4.33 that $f(S_g\omega) = h(g/G_m)$ for $g/G_m \in K$, i.e. 2.4.31 holds.

<u>Theorem 2.4.11</u> (Batty, Bratteli - Elliott - Robinson)

Let $A = C_0(\Omega)$ be an abelian C^*-algebra, G a Lie group, $\sigma: G \to \mathrm{Aut}(A)$ a strongly continous action of G on A, $\Delta = \{\delta_1 \cdots \delta_d\}$ a basis for the corresponding action of $\mathfrak{g}$ = the Lie algebra of G on A, and let A_n be the corresponding algebra of n times differentiable elements, $n = 0, 1, 2, \ldots, \infty$. If δ is a derivation in $\mathrm{Der}(A_n, A)$, then there exists real functions $\ell_1 \ldots \ell_d$ on Ω such that

$$(2.3.34)\qquad (\delta f)(\omega) = \sum_{k=1}^{d} \ell_k(\omega)(\delta_k f)(\omega)$$

for all $f \in A_n$, $\omega \in \Omega$.

<u>Proof</u>

We partly follow [Bat 4], [BER 1] and [BaR 2].

The case $n = \infty$ follows by trivial modifications of the proof in the case that n is finite, so assume that n is finite. By theorem 2.3.1, there exists a constant $C>0$ such that

$$(2.4.35)\quad |\delta f(\omega)| \le C\{\|f\| + \sum_{k=1}^{n} \sum_{i_1=1}^{d} \cdots \sum_{i_k=1}^{d} \|\delta_{i_1} \cdots \delta_{i_k} f\| / k!\} .$$

for all $\omega \in \Omega$. If f vanishes in a neighbourhood of ω, then by lemma 2.4.5, $(\delta f)(\omega) = 0$. But if f is in A_{2n} and $(\delta_{i_1} \cdots \delta_{i_k} f)(\omega) = 0$ for $0 \le k \le n$, $i_j \in \{1, \ldots d\}$, then f can be approximated in $\| \ \|_n$-norm by functions in A_n vanishing in a neighborhood of ω by Lemma 2.4.6. It follows from the estimate 2.4.35 that $(\delta f)(\omega)=0$. Therefore, the intersection of kernels of the functionals

$$A_{2n} \ni f \to (\delta_{i_1}\cdots\delta_{i_k} f)(\omega) \ , \quad 0\le k\le n \ , \ i\in \{1,\dots d,\},$$

is contained in the kernel of

$$A_{2n} \ni f \to (\delta f)(\omega).$$

Hence, by linear algebra, the latter functional is a linear combination of the former, that is, there exists numbers $\ell_{i_1,\dots,i_k}(\omega)$ such that

$$(2.4.36) \quad (\delta f)=\ell_{\phi}(\omega)f(\omega)+\sum_{k=1}^{n}\ \sum_{i_1=1}^{d}\cdots\sum_{i_k=1}^{d}\ell_{i_1,\dots,i_k}(\omega)(\delta_{i_1}\cdots\delta_{i_k}f)(\omega)$$

for all $f\in A_{2n}$. But the left side of 2.4.36 is continuous as a functional on $f\in A_n$ by 2.4.35, while the right side trivially is continuous, for fixed ω. It follows from Lemma 2.4.9 that the relation 2.4.36 extends by continuity to all $f\in A_n$.

It remains to show that we can choose the coefficients $\ell(\omega)$ such that

$$\ell_{\phi}(\omega)=0 \ , \ \ell_{i_1,\dots,i_k}(\omega)=0 \text{ for } k=2,\ 3,\ \dots$$

To this end, let S be the G-flow on Ω corresponding to σ by 2.4.1, and let G_ω be the stabilizer group at ω, see 2.4.30. Now choose a basis $X_1, \dots , X_d$ for the Lie algebra of G such that there is an m such that $X_{m+1}, \dots , X_d$ is a basis for the Lie algebra of the closed subgroup G_ω, and redefine δ_i as

$$\delta_i = d\sigma(X_i),$$

see 2.4.22. This only amounts to a linear rearrangement of each of the k-homogeneous terms $\sum_{i_1=1}^{d}\cdots\sum_{i_k=1}^{d}\ell_{i_1,\dots,i_k}(\omega)(\delta_{i_1}\cdots\delta_{i_k}f)(\omega)$ in 2.4.36. But then $(\delta_{i_1}\cdots\delta_{i_k}f)(\omega)=0$ if one of the digits $i_1, \dots , i_k$ is contained in $\{m+1, \dots , d\}$. Therefore we can put $\ell_{i_1,\dots,i_k}(\omega)=0$ for such sequences $(i_1, \dots, i_k)$, and 2.4.36 reduces to

$$(2.4.37) \quad (\delta f)(\omega)=\ell_{\phi}(\omega)f(\omega)+\sum_{k=1}^{n}\ \sum_{i_1=1}^{m}\cdots\sum_{i_k=1}^{m}\ell_{i_1,\dots,i_k}(\omega)(\delta_{i_1}\cdots\delta_{i_k}f)(\omega)$$

But using the Lie algebra relations:

$$\delta_i\delta_j-\delta_j\delta_i = d\sigma([X_i, X_j])$$
$$= \text{linear combination of } \delta_k \text{ for } k=1, \ldots, d,$$

we can further rearrange 2.4.37 and redefine ℓ so it has the form

(2.4.38) $(\delta f)(\omega)=$

$$\ell_\phi(\omega)f(\omega)+\sum_{k=1}^{n}\sum_{1\le i_1\le i_2\le\ldots i_k\le m}\ell_{i_1,\ldots,i_k}(\omega)(\delta_{i_1}\ldots\delta_{i_k}f)(\omega),$$

and it is a straightforward consequence of Lemma 2.4.10 that the coefficients ℓ in 2.4.38 are unique.

We now use the derivation property of δ to argue that $\ell_{i_1,\ldots,i_k}(\omega)=0$ if $k\ge 2$. By Lemma 2.4.10, for all $(x_1, \ldots, x_m)\in \mathbb{R}^m$, there exists an $f\in C_0^\infty(\Omega)$ such that

$f(\omega)=1$, $(\delta_i f)(\omega)=x_i$, $i=1, \ldots, m)$ and

$(\delta_{i_1} \ldots \delta_{i_k}f)(\omega)=0$, $k\ge 2$

(Use that G/G_ω is locally homomorphic to the unit ball in $\mathbb{R}^m$). Then

$(\delta_{i_1}\ldots\delta_{i_k} f^k)(\omega)=k^j x_{i_1}\ldots x_{i_j}+O(k^{j-1})$ but

$(\delta f^k)(\omega)=kf^{k-1}(\omega)(\delta f)(\omega)=O(k)$,

for $k=1, 2, 3, \ldots$. Now, comparing orders on both sides of 2.4.38, and letting $(x_1\ldots x_m)$ vary in $\mathbb{R}^m$, we have

$$\ell_{i_1,\ldots,i_j}(\omega)=0$$

whenever $j\ge 2$.. Thus

(2.4.39) $(\delta f)(\omega)=\ell_\phi(\omega)f(\omega)+\sum_{i=1}^{m}\ell_i(\omega)(\delta_i f)(\omega)$

for all $f\in A_n$. But if $f, g\in A_n$ it follows from 2.4.39 and the derivation property of δ and δ_i, $i=1, \ldots, m$ that

$$\ell_\phi(\omega)f(\omega)g(\omega)=0$$

and this finally establishes that $\ell_\phi(\omega)=0$, and the expansion 2.4.34 of the theorem is valid.

Remark 2.4.12

Note that the proof of the expansion 2.4.36 only depended on the estimate 2.4.35 and the fact that δ is local in the sense that

(2.4.40) $\quad \mathrm{supp}\,(\delta(f)) \subseteq \mathrm{supp}\,(f)$

for all $f \in A_n$. By using a technique somewhat reminiscent of the proof of Theorem 2.3.1, one can show that any linear operator δ from A_n into A which is local in the sense of 2.4.40 is continuous in the Fréchet topology on A_n, and thus has the expansion 2.4.36, see [BER 1] and [BaR 2].

Problem 2.4.13

Let A be an abelian C^*-algebra, $\Delta = \{\delta_1, \ldots, \delta_d\}$ a finite collection of closed derivations on A, and assume that A_n is dense for some $n = 1, 2, \ldots, \infty$. Does any derivation $\delta \in \mathrm{Der}\,(A_n, A)$ have an expansion

$$(\delta f)(\omega) = \sum_{k=1}^{d} \ell_k(\omega)(\delta_k f)(\omega)$$

for all $\omega \in \Omega$ and $f \in A_n$ (or $f \in A_{2n}$)? Note that δ has the expansion 2.4.36, by the proof of theorem 2.4.11.

In Theorem 2.4.11, nothing is stated about continuity and other properties of the functions ℓ_i. This is a difficult problem to analyze in general, and the difficulty stems from the fact that the stabilizer subgroups $G_\omega \subseteq G$ varies with ω. Particularly nasty situations occur if the Lie group dimension of G_ω is positive, because then the coefficients $\ell_i(\omega)$ are not even all unique. These types of degeneracies can be avoided in the following cases.

Definition 2.4.14

Let G be a locally compact group, and S a G-flow on a locally compact Hausdorff space Ω. For each $\omega \in \Omega$, let $G_\omega = \{g \in G \mid S_g\omega = \omega\}$ be the stabilizer group at ω.

The flow S is said to be free if $G_\omega = \{e\}$ for all $\omega \in \Omega$. The flow is locally free if for all $\omega \in \Omega$ there exists a neighborhood V_ω of e in G such that $G_\omega \cap V_\omega = \{e\}$ (i.e. G_ω is a discrete subgroup of G). The flow is uniformly locally free if there exists a neighborhood V of e in G such that $G_\omega \cap V = \{e\}$ for all $\omega \in \Omega$.

If σ is the associated action of G on $C_0(\Omega)$, σ is said the be free resp. locally free, resp. uniformly locally free if S has any of these three properties, respectively.

Theorem 2.4.15

Let $A = C_0(\Omega)$ be an abelian C*-algebra, G a Lie group, $\sigma: G \to \mathrm{Aut}(A)$ a strongly continuous action of G on A, and $\Delta = \{\delta_1, \ldots, \delta_d\}$ a basis for the corresponding action of $\mathfrak{g}$. If the action σ is locally free, then for any derivation $\delta \in \mathrm{Der}(A_n, A)$ there exists unique functions $\ell_1, \ldots, \ell_d$ on Ω such that

$$(2.4.41)\quad (\delta f)(\omega) = \sum_{k=1}^{d} \ell_k(\omega)(\delta_k f)(\omega)$$

for all $f \in A_n$, $\omega \in \Omega$, and these functions are real and continuous. If the action σ is uniformly locally free, the functions ℓ_k are uniformly bounded.

Remark 2.4.16

If conversely the functions ℓ_k are continuous and uniformly bounded, then the formula 2.4.21 defines a derivation $\delta \in \mathrm{Der}(A_1, A)$, so in particular any derivation in $\mathrm{Der}(A_\infty, A)$ extends uniquely to a derivation in $\mathrm{Der}(A_1, A)$, and the classification of the spaces $\mathrm{Der}(A_n, A)$ is complete in this case.

If the action σ is only locally free, one can not conclude boundedness of the coefficients in (2.4.41). One counterexample is $G = \mathbb{R}$, $\Omega = \mathbb{T} \times \mathbb{R} = \{z \in \mathbb{C} \mid |z| = 1\} \times \mathbb{R}$, and the flow S is defined by $S_t(z, x) = (e^{itx} z, x)$. If $\delta_0 = -ix \frac{\partial}{\partial\theta}$ is the generator of S, then the derivation δ defined by $(\delta f)(z,x) = \ell(z,x)(\delta_0 f)(z,x)$ is in $\mathrm{Der}(A_\infty, A)$ if and only if ℓ is a continuous real function such that there exists constant C and n with $|\ell(z,x)| \le C(1 + |x|^n)$ for all z,x. This is clear from Theorem 2.4.17.

Proof of Theorem 2.4.15.

The existence of ℓ_k is a consequence of Theorem 2.4.11, and the uniqueness is clear from the last part of the proof of Theorem 2.4.11. Fix an $\omega_0 \in \Omega$. As the action is locally free, G_{ω_0} is a discrete subgroup of G, and it follows from Lemma 2.4.10 that there exists functions $f_k \in C_{00}^{\infty}(\Omega)$ such that

$$(2.4.42) \quad (\delta_i f_k)(\omega_0) = \delta_{ik}, \qquad i, k = 1, \ldots, d.$$

But

$$\sum_{i=1}^{d} \ell_i(\omega)(\delta_i f_k)(\omega) = \delta(f_k)(\omega) \quad \text{for} \quad k = 1, \ldots, d$$

constitutes a set of d linear equations for $\ell_i(\omega)$, and as all coefficients of these equations depends continuously of ω, and as the determinant is equal to 1 at ω_0 by 2.4.42, it follows that the solution $\ell_1(\omega), \ldots, \ell_d(\omega)$ is continuous in a neighborhood of ω_0. Thus $\ell_1, \ldots, \ell_d$ are locally continuous functions, and hence they are continuous functions.

We next prove that the coefficient functions ℓ_i are uniformly bounded when the flow is uniformly locally free. Let V be a neighborhood of e in G such that $\overline{V}$ is compact and $V \cap G_\omega = \{e\}$ for all $\omega \in \Omega$, and let W be another neighborhood of e such that $W^4 \subseteq V$ and $W = W^{-1}$. Assume ad absurdum that one of the coefficient functions, say ℓ_1, is unbounded, and let $\omega_n \in \Omega$ be a sequence such that $|\ell_1(\omega_n)| \to \infty$ as $n \to \infty$. Then $\omega_n \to \infty$, and by thinning out the sequence we may assume that each set $\overline{S_{W^3}\omega_n}$ has a compact neighborhood K_n in Ω such that the compact sets $\overline{S_W K_n}$ are disjoint for different n, and we may also assume $\ell_1(\omega_n) \neq 0$ for all n. Let $h \in C_0^{\infty}(G)$ be a function such that $(X_1 h)(e) = 1$ and $(X_i h)(e) = 0$ for $i = e, \ldots, d$ where $X_i \in \mathfrak{g}$ is the element such that $\delta_i = d\sigma(X_i)$. Using Theorem 2.2.3 as in the proof of Lemma 2.4.10, there exists functions $\psi_i \in C_{00}^{\infty}(G)$, $h_i \in C_{00}(G)$ such that $\operatorname{supp} \psi_i \subseteq W$, $\operatorname{supp} h_i \subseteq W^3$, and $h(g) = \sum_{i=1}^{m} \int_G dg' \, \psi_i(g') \, h_i(g'^{-1}g)$ for $g \in W$. Now, as the maps $g \in \overline{W^3} \to S_g\omega_n$ are homeomorphisms for each n, the functions h_i can be viewed as functions $h_{i,n}$ on $S_{\overline{W^3}}\omega_n$ by $h_{i,n}(S_g\omega_n) = h_i(g)$, $g \in \overline{W^3}$.

Let $h_{i,n}$ also denote an arbitrary continuous extension of $h_{i,n}$ to a function in $C_{00}(\Omega)$ with support in K_n and such that $\| h_{i,n}\| \leq \|h_i\|$, and define $f_n = \sum_{i=1}^{m} \int dg\, \psi_i(g)\sigma_g(h_{i,n})$. Then $f_n \in C_{00}^{\infty}(\Omega)$, supp $f_n \subsetneq \overline{S_W K_n}$ and $(\delta_1 f_n)(\omega_n) = 1$, $(\delta_i f_n)(\omega_n) = 0$, $i = 2, \dots, d$. But by partial integration, $\delta_{i_1} \cdots \delta_{i_k} f_n$

$= d\sigma(X_{i_1}) \dots d\sigma(X_{i_k}) f_n = \sum_{i=1}^{m} \int_G dg((-X_{i_k}) \dots (-X_{i_1})\psi_i)(g)\sigma_g(h_{i,n})$

and as the functions ψ_i are independent of n, and $\|h_{i,n}\| = \|h_i\|$ is independent of n, we get estimates on $\|\delta_{i_1} \cdots \delta_{i_k} f_n\|$ which are uniform in n. Since the functions f_n has disjoint and compact supports, it follows from Lemma 2.4.5 that the function

$$f = \sum_n \frac{1}{|\ell(\omega_n)|^{\frac{1}{2}}} f_n \text{ is contained in } A_\infty. \text{ But } (\delta f)(\omega_n) = \frac{\ell_1(\omega_n)}{|\ell_1(\omega_n)|^{\frac{1}{2}}} \to \infty$$

as $n \to \infty$, and this contradicts $\delta f \in C_0(\Omega)$. Thus the coefficient functions ℓ_i must be bounded.

In the particular case $G = \mathbb{R}$, the stabilizer subgroups G_ω are particularly easy to classify, they depend on only one parameter: the frequency of ω.

Definition 2.4.17

Let S be an ($\mathbb{R}$-) flow on Ω. If $\omega \in \Omega$, the period of ω is defined as

$$(2.4.43) \quad p(\omega) = \inf\{t > 0 \mid S_t\omega = \omega\},$$

with the convention that $p(\omega) = +\infty$ if $S_t\omega \neq \omega$ for all $t \neq 0$. The frequency of ω is defined as

$$(2.4.44) \quad \nu(\omega) = 1/p(\omega)$$

with the convention that $\nu(\omega) = 0$ if $p(\omega) = +\infty$ and $\nu(\omega) = +\infty$ if ω is a fixed point for the flow, i.e. if $p(\omega) = 0$.

Note that the connection between the frequency $\nu(\omega)$ and the stabilizer subgroup G_ω is that $G_\omega = \nu(\omega)^{-1}\mathbb{Z} = p(\omega)\mathbb{Z}$, with the convention that $p(\omega)\mathbb{Z} = \mathbb{R}$ if $p(\omega) = 0$ and $p(\omega)\mathbb{Z} = \{0\}$ if $p(\omega) = +\infty$. Thus, in the case $0 < \nu(\omega) < \infty$, the orbit

(2.4.45) $S_{\mathbb{R}}\omega = \{S_t\omega \mid t \in \mathbb{R}\}$.

of ω in Ω is homeomorphic to the circle $\mathbb{T}$, if $\nu(\omega) = \infty$, then $S_{\mathbb{R}}\omega = \{\omega\}$ and if $\nu(\omega) = 0$, then $S_{\mathbb{R}}\omega$ is locally homeomorphic to the line $\mathbb{R}$ in the sense that $S_{[-N,N]}$ is homeomorphic to $[-N,N]$ for all $N > 0$. Simple examples, like the irrational flows on the two-torus $\mathbb{T}^2$, show that $S_{\mathbb{R}}\omega$ is not necessarily globally homeomorphic with $\mathbb{R}$.

Note that a flow is free if and only if $\nu(\omega) = 0$ everywhere, locally free if and only if $\nu(\omega) < +\infty$ everywhere, i.e. if and only if there are no fixed points, and uniformly locally free if and only if $\sup\{\nu(\omega) \mid \omega \in \Omega\} < +\infty$.

The elementary continuity properties of $\omega \to \nu(\omega)$ is summarized in the following Lemma:

Lemma 2.4.18

The map $\omega \mapsto \nu(\omega)$ is upper semicontinuous, i.e. if $\omega_\alpha \to \omega$ then
(2.4.46) $\nu(\omega) \geq \overline{\lim_\alpha}\, \nu(\omega_\alpha)$.

Furthermore, if $\nu(\omega)<\infty$ and $\bar{\nu} = \overline{\lim_\alpha}\, \nu(\omega_\alpha)>0$ then

(2.4.47) $\frac{\nu(\omega)}{\bar{\nu}} \in \mathbb{N} = \{1, 2, 3, \ldots\}$.

Remark 2.4.19

Simple examples, like doubling an 8 into an 0 (i.e. the standard flow on the Möbius band) or the flow generated by $y\frac{d}{dx}$ on $\mathbb{R}^2$, show that $\omega \to \nu(\omega)$ is not continuous in general.

Proof of Lemma 2.4.18. If $p = \underline{\lim_\alpha}\, p(\omega_\alpha)$, it is enough to show that $p(\omega) \leq p$, and that $p/p(\omega)$ is an integer provided $p(\omega)>0$ and $p<+\infty$. If $p=+\infty$ these statements are trivial, if $p<+\infty$ it suffices to show that $S_p\omega = \omega$. We may assume that $\lim_\alpha p(\omega_\alpha) = p$ exists by passing to a subnet. But as

$$S_{p(\omega_\alpha)}\,\omega_\alpha = \omega_\alpha$$

for all α , it follows by limiting and by joint continuity of S that $S_p\,\omega = \omega$.

In the case of flow, we can give a satisfactory completion of Theorem 2.4.11:

Theorem 2.4.20 (Bratteli-Elliott-Robinson)

Let $A = C_0(\Omega)$ be an abelian C^*-algebra, $\sigma:\mathbb{R} \mapsto \mathrm{Aut}(A)$ a strongly continuous action of $\mathbb{R}$ on A with associated flow S and generator δ_0 . Let $\Omega_0 \subsetneq \Omega$ be the set of fixed points in S. We say that a function λ on $\Omega\backslash\Omega_0$ is differentiable if

$$(\delta_0\lambda)(\omega) \equiv \lim_{t\to 0} \frac{1}{t}(\lambda(S_t\omega) - \lambda(\omega))$$

exists for all $\omega \in \Omega\backslash\Omega_0$ and $\omega \mapsto (\delta_0\lambda)(\omega)$ is continuous on $\Omega\backslash\Omega_0$. Let $n, m \in \{0, 1, 2, \ldots, \infty\}$.

The following conditions are equivalent.

1. $\delta \in \mathrm{Der}(A_n, A_m)$.

2. a. If $n<m+1$, then $\delta = 0$.

   b. If $n\geq m+1$, then there exists a continuous function ℓ on $\Omega\backslash\Omega_0$ such that

$$(\delta f)(\omega) = \begin{cases} \ell(\omega)(\delta_0 f)(\omega) & \text{for } \omega \in \Omega\backslash\Omega_0 \\ 0 & \text{for } \omega \in \Omega_0 . \end{cases} \tag{2.4.48}$$

   The function ℓ is m times differentiable and:

   b.1. If $n = \infty$, the derivatives $\delta_0^k\ell$ are polynomially bounded in the frequency for each $k < m+1$, i.e. there exists for each $k < m+1$ constants C, p such that

$$|(\delta_0^k\ell)(\omega)| \leq C(1+\nu(\omega)^p) \tag{2.4.49}$$

   for all $\omega \in \Omega\backslash\Omega_0$.

b.2. If $n < \infty$, there exists a constant $C > 0$ such that

(2.4.50) $$|(\delta_o^k \ell)(\omega)| \le C(1+\nu(\omega)^{n-m+k-1}) \quad \text{for } k = 0, 1, \ldots, m$$

Proof

The analogue of this theorem for local operators, 2.4.40, is proved in detail in [BER 1]. In proving $2 \Rightarrow 1$ one uses the fact that if $f \in A_n$ then the function.

(2.4.51) $$\omega \in \Omega \backslash \Omega_o \mapsto \nu(\omega)^q (\delta_o^p f)(\omega)$$

vanishes at infinity provided $q + p \le n$ and $p \ge 1$. Since $\delta_o(A_n) \subsetneq A_{n-1}$ for all n it suffices to show this for $p = 1$. So assume $p = 1$ and $k = 0, 1, \ldots, n-1$, and let ω_i be a net of points in $X\backslash X_o$ converging to ∞ in $X\backslash X_o$. We must show that $\nu(\omega_i)^k (\delta_o f)(\omega_i)$ converges to zero. We may distinguish between two cases (by passing to a subnet if necessary). (Either Case 1 occurs for a subnet, or Case 2 occurs for the whole net.)

Case 1 $\nu(\omega_i) \le 1$ for all i.

Since $\delta_o f \in A$, and $(\delta_o f)(\omega) = 0$ for $\omega \in \Omega_o$, $\delta_o f$ vanishes at ∞ as a function on $\Omega\backslash\Omega_o$. As $|\nu(\omega_i)^k| \le 1$ for all i, it follows that

$$\lim_{i\to\infty} \nu(\omega_i)^k (\delta_o f)(\omega_i) = 0$$

Case 2 $\nu(\omega_i) \ge 1$ for all i.

It is sufficient to consider the case that f is real-valued. The functions

$$t \in \mathbb{R} \to (\delta_o f)(S_t \omega_i) \equiv g_i(t)$$

are periodic with period $p_i = \frac{1}{\nu(\omega_i)} \le 1$. Also, as g_i is the derivative of the function $t \to f(S_t\omega_i)$ which is also periodic with period p_i, and real-valued, there is a $t_o \in [0,1]$ such that $g_i(t_o) = 0$. But then for an arbitrary t we have

$$|g_i(t)| \le |g_i(t_o)| + |t-t_o| \, \|g_i'\| = |t-t_o| \, \|g_i'\|$$

by the mean value theorem, and since g_i is periodic with period p_i we find

$$|g_i(t)| \le \frac{p_i}{2} \|g_i'\| \le p_i \|g_i'\| .$$

As $f \in A_n$ we have $g_i \in C^n(R)$ and iterating the argument above we obtain

$$\|g_i\| \le p_i^n \|g_i^{(n)}\| = \frac{1}{\nu(\omega_i)^n} \|g_i^{(n)}\| ,$$

and hence

$$|\nu(\omega_i)^k(\delta f)(\omega_i)| \le \frac{1}{\nu(\omega_i)^{n-k-1}} \|\delta^n f|_{T_{[0,1]}\omega_i}\| .$$

Now, as ω_i converges to infinity, so does $T_{[0,1]}\omega_i$, and it follows that

$$\lim_{i\to\infty} \nu(\omega_i)^k (\delta f)(\omega_i) = 0 .$$

for $k = 0, 1, \ldots, n-1$. This ends the proof of 2.4.51, and thus of $2 \Rightarrow 1$. In proving $1 \Rightarrow 2$ one uses Theorem 2.4.11 to get the existence of ℓ, then the method of proof in Theorem 2.4.15 to show continuity on the open set $\Omega\backslash\Omega_0$. In the case that $\delta \in \mathrm{Der}(A_n, A)$ one use a similar technique as in the proof of the last part of Theorem 2.4.15 to construct functions $f \in A_n$ which optimizes the estimate 2.4.51 to get the estimates 2.4.49 and 2.4.50. If $\delta \in \mathrm{Der}(A_n, A_m)$ one uses that the operators defined by

$$\delta_0^k(\ell(\delta_0 f))(\omega) = \sum_{q=0}^{k} \binom{k}{q} \delta_0^q \ell(\omega) \delta_0^{k-q+1} f(\omega)$$

are local operators from A_n into A for all $k < m+1$.

Problem 2.4.21

Prove an analogue of Theorem 2.4.20 for an action of a general Lie group G instead of $\mathbb{R}$. In that case the subgroups G_ω has to be specified by more parameters than only the frequency ν. If $G = \mathbb{R}^d$, one can for example use a basis for the linear subspace $\mathfrak{g}_\omega$(=the Lie algebra of G_ω) together with a finite set $\{g_1,\ldots,g_m\} \subseteq G_\omega$ such that the image of $\{g_1,\ldots,g_m\}$ in $G/\exp\{\mathfrak{g}_\omega\}$ is linearly independent, and such that $\exp\{\mathfrak{g}_\omega\}$ together with $\{g_1,\ldots,g_m\}$ generate G_ω as a group. See the remarks after Theorem 1.1 in [BER 1].

§ 2.4.2 Generators of flows

To aid the intuition, we start discussing flows on $\mathbb{R}$ with generators of the form $\ell(x)\frac{d}{dx}$, where ℓ is a continuous function and $\frac{d}{dx}$ is the usual derivative. Then $\ell(x)$ can be interpreted as the velocity of the flow at the point x, and if $[a,b] \subseteq \mathbb{R}$, then

$$(2.4.52) \quad T(a,b) = \int_a^b \frac{dx}{\ell(x)}$$

can be interpreted as the time a point following the flow uses from point a to point b. We see immediately that problems occur at the points x where $\ell(x) = 0$. For example, if $\ell(0) = 0$, but $\ell(x) > 0$ if $0 < |x| \leqslant \varepsilon$ for some $\varepsilon > 0$ and $\int_{-\varepsilon}^{0} \frac{dx}{\ell(x)} < +\infty$, but $\int_0^{\varepsilon} \frac{dx}{\ell(x)} = +\infty$, then points in the interval $[-\varepsilon, 0\rangle$ reaches the point 0 is a finite time, but the point 0 uses an infinite amount of time to reach any point in $\langle 0, \varepsilon]$, thus $\ell\frac{d}{dx}$ cannot generate a one-parameter group of homeomorphisms. One way to avoid these degeneracies is to assume that $1/\ell$ is nonintegrable to the left and right of any point x where $\ell(x) = 0$, then these points will be fixed points for the generated flow. There is a standard assumption of ℓ assuming this, namely that ℓ is Lipschitz continuous, and the following lemma is the standard existence and uniqueness theorem for solutions of first order ordinary differential equations.

Lemma 2.4.22

Let $\ell: \mathbb{R} \to \mathbb{R}$ be a function which is uniformly Lipschitz continuous in the sense that there exists a constatnt $K > 0$ such that

$$(2.4.53) \quad |\ell(x) - \ell(y)| \leqslant K|x - y|$$

for all $x, y \in \mathbb{R}$. Then the initial value problem

$$(2.4.54) \quad \begin{aligned} \frac{d}{dx}f(x) &= \ell(f(x)) \\ f(0) &= y \end{aligned}$$

has a unique solution $f_y: \mathbb{R} \to \mathbb{R}$ for each $y \in \mathbb{R}$. These solutions satisfy the group property

(2.4.55) $f_y(t + s) = f_{f_y(t)}(s)$

and hence if we define

(2.4.56) $T_t y = f_y(t)$

for $y, t \in \mathbb{R}$, then T is a flow on $\mathbb{R}$.

Proof

Local existence and uniqueness of the solution follows by the usual method of successive approximation, global existence follows from the uniformity of K and the group property follows as ℓ only depends explicitly on the dependent variable f. That T is a flow, especially that each T_t is a homeomorphism, follows from stability theory for ordinary differential equations; this is essentially the proof of Case 1 in the argument for Theorem 2.4.24, see below.

When ℓ is not Lipschitz continuous, it may happen that $\ell\frac{d}{dx}$ has no generator extensions, as mentioned before Lemma 2.4.22, or it may happen that $\ell\frac{d}{dx}$ has a continuum of generator extensions:

Example 2.4.23 (P.O. Frederickson)

Let ℓ be a nonnegative continuous bounded function on $\mathbb{R}$ such that the set $C = \{x \in \mathbb{R} \mid \ell(x) = 0\}$ is the Cantor subset of $[0, 1]$, and such that $1/\ell$ is Lebesque integrable over $[-1, 2]$:

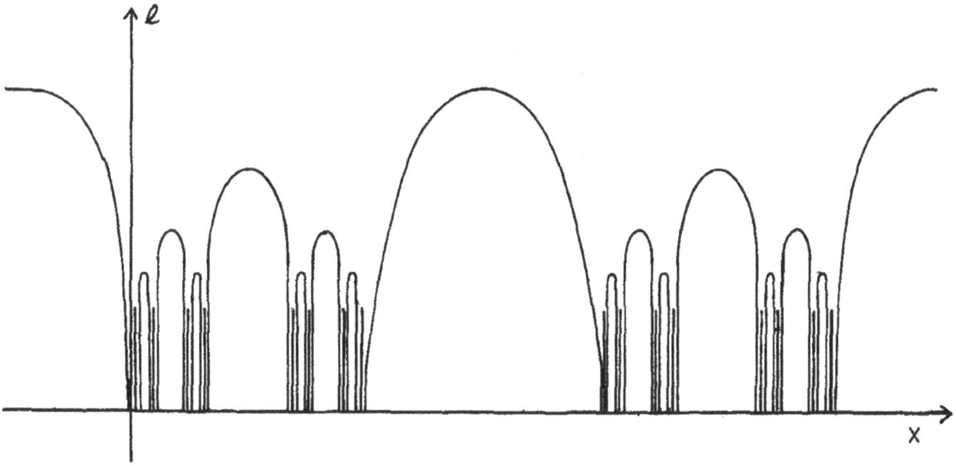

Then, if μ is any positive non-atomic measure on C, we can define a flow T on $\mathbb{R}$ by requiring that the time a point following the flow uses from a to b is

$$T(a, b) = \int_a^b \frac{dx}{\ell(x)} + \mu([a, b] \cap C)$$

The generator of each of these flows is an extension of $\ell\frac{d}{dx}$. For a more systematic study in this direction, see [Bat 5], [Bat 6], [Kur 3]. Using the notation of Theorem 1.6.20, it is proved in the latter paper that a closed derivation δ on $C(I)$ is a generator if and only if

(i) $A = B$ and A is an open subset of $(0, 1)$.

(ii) The restriction $\mu|_C$ to any connected component C of A is a homeomorphism from C onto $\mathbb{R}$.

(iii) $\delta = \delta(A, A, \mu)$.

One way to ensure that ℓ is uniformly Lipschitz continuous is to require that ℓ is differentiable with bounded derivative. Combining with Theorem 2.4.20, this motivates the following theorem.

<u>Theorem 2.4.24</u> (Bratteli-Digernes-Goodman-Robinson)

Let $\mathcal{A} = C_0(\Omega)$ be an abelian C*-algebra, $\sigma: \mathbb{R} \to \mathrm{Aut}\,(\mathcal{A})$ a strongly continuous action of $\mathbb{R}$ on $\mathcal{A}$ with assoicated flow S, and let $\delta \in \mathrm{Der}\,(\mathcal{A}_\infty, \mathcal{A}_1)$.

It follows that δ is a pre-generator.

Furthermore, if δ_0 is the generator of σ, there is a continuous function ℓ on $\Omega\setminus\Omega_0$ such that $\delta = \ell\delta_0$ in the sense of 2.4.48. Here Ω_0 is the set of fixed points for S. If $\omega \in \Omega\setminus\Omega_0$, define the function $\ell_\omega: \mathbb{R} \to \mathbb{R}$ by

(2.4.57) $\ell_\omega(t) = \ell(S_t\omega)$.

Then the initial value problem

(2.4.58) $$x_\omega'(t) = \ell_\omega(x_\omega(t)), \quad t \in \mathbb{R}, \qquad x_\omega(0) = 0$$

has a unique solution $t \in \mathbb{R} \to x_\omega(t)$, and if we define

$$(2.4.59)\quad T_t = \begin{cases} S_{x_\omega(t)}\omega & \text{for } \omega \in \Omega\setminus\Omega_0 , \\ \omega & \text{for } \omega \in \Omega_0 , \end{cases}$$

for all $t \in \mathbb{R}$, then T is the flow generated by the closure of δ.

Proof

This theorem is proved in detail in [BDGR 1]. The proof of the existence of a flow T such that the generator δ_T of T extends δ is relatively straightforward, while the proof of uniqueness of T and that δ_T is the closure of δ is somewhat long, and we refer to [BDGR 1] for the details of latter proof.

First, Theorem 2.4.20 implies that δ has the form

$$(2.4.60)\quad (\delta f)(\omega) = \begin{cases} \ell(\omega)(\delta_0 f)(\omega) & \text{for } \omega \in \Omega\setminus\Omega_0 , \\ 0 & \omega \in \Omega_0 , \end{cases}$$

where ℓ is a once differentiable continuous function on $\Omega\setminus\Omega_0$ such that ℓ and $\delta_0\ell$ is polynomially bounded in the frequency, i.e.

$$(2.4.61)\quad |\ell(\omega)| \leqslant K(\nu(\omega))$$

$$(2.4.62)\quad |(\delta_0\ell)(\omega)| \leqslant K(\nu(\omega))$$

where K is a polynomial. It follows from 2.4.62 that ℓ is Lipschitz continuous an orbits, i.e.

$$(2.4.63)\quad |\ell(S_t\omega) - \ell(S_s\omega)| \leqslant K(\nu(\omega))|t - s| .$$

Here we used the fundamental theorem of calculus and the fact that ν is constant an orbits. From 2.4.57 and 2.4.63 it follows that

$$(2.4.64)\quad |\ell_\omega(t) - \ell_\omega(s)| \leqslant K(\nu(\omega))|t - s|$$

and hence Lemma 2.4.22 implies that 2.4.58 has a unique solution, and 2.4.59 defines a one-parameter group $t \in \mathbb{R} \to T_t$ of bijective maps on Ω.

We must show that T is a flow in the sense of Definition 2.4.2, i.e. that $(t, \omega) \in \mathbb{R}\times\Omega \to T_t\omega \in \Omega$ is jointly continuous.

To this end, assume that $(t_\alpha, \omega_\alpha) \to (t, \omega)$ in $\mathbb{R} \times \Omega$. We have to show that $T_{t_\alpha}\omega_\alpha \to T_t\omega$ in Ω, i.e.

(2.4.65) $S_{x_{\omega_\alpha}(t_\alpha)}\omega_\alpha \to S_{x_\omega(t)}\omega$

We divide the discussion into two cases

Case 1 $\nu(\omega) < \infty$

Since S is jointly continuous, it suffices to show

(2.4.66) $x_{\omega_\alpha}(t_\alpha) \to x_\omega(t)$

But as

$$|x_\omega(t) - x_{\omega_\alpha}(t_\alpha)| \leq |x_\omega(t) - x_\omega(t_\alpha)| + |x_\omega(t_\alpha) - x_{\omega_\alpha}(t_\alpha)|$$

and $x_\omega(t_\alpha) \to x_\omega(t)$, it suffices to show that

(2.4.67) $x_{\omega_\alpha}(s) - x_\omega(s) \to 0$

uniformly for s in compacts. We show uniform convergence on intervals of the form $[0, t_0]$, the case $[-t_0, 0]$ follows by trivial modifications.

It follows from 2.4.58 that

(2.4.68) $|x_{\omega_\alpha}(t) - x_\omega(t)| \leq \int_0^t ds|\ell_{\omega_\alpha}(x_{\omega_\alpha}(s)) - \ell_\omega(x_\omega(s))|$

for $t > 0$. We are assuming $\nu(\omega) < \infty$, and as $\nu(\omega) = \overline{\lim_\alpha}\, \nu(\omega_\alpha)$ by Lemma 2.4.18, it follows that the frequencies $\nu(\omega_\alpha)$ are uniformly bounded. Hence it follows from 2.4.61 and 2.4.63 that there exists a finite constant K such that

(2.4.69) $|\ell_{\omega_\alpha}(x) - \ell_{\omega_\alpha}(y)| \leq K|x - y|$
$|\ell_\omega(x) - \ell_\omega(y)| \leq K|x - y|$

(2.4.70) $|\ell_{\omega_\alpha}(x)| \leq K$
$|\ell_\omega(x)| \leq K$

From 2.4.58 and 2.4.70 we have

(2.4.71) $|x_{\omega_\alpha}(t)| \leqslant Kt$,

and hence 2.4.69 and 2.4.71 implies:

(2.4.72)
$$\begin{aligned}
&|\ell_{\omega_\alpha}(x_{\omega_\alpha}(s)) - \ell_\omega(x_\omega(s))| \\
&\leqslant |\ell_{\omega_\alpha}(x_{\omega_\alpha}(s)) - \ell_{\omega_\alpha}(x_\omega(s))| + |\ell_{\omega_\alpha}(x_\omega(s)) - \ell_\omega(x_\omega(s))| \\
&\leqslant K|x_{\omega_\alpha}(s) - x_\omega(s)| + M_\alpha
\end{aligned}$$

for $0 \leqslant s \leqslant t_0$, where

(2.4.73) $M_\alpha = \sup\{|\ell_{\omega_\alpha}(x) - \ell_\omega(x)| \mid |x| < Kt_0\}$

Now, 2.4.68 and 2.4.70 gives a crude first estimate

$$|x_{\omega_\alpha}(t) - x_\omega(t)| \leqslant 2Kt$$

This inserted in 2.4.72 gives

$$|\ell_{\omega_\alpha}(x_{\omega_\alpha}(s)) - \ell_\omega(x_\omega(s))| \leqslant 2K^2s + M_\alpha$$

This inserted in 2.4.68 gives the better estimate

$$|x_{\omega_\alpha}(t) - x_\omega(t)| \leqslant K^2t^2 + M_\alpha t$$

and this inserted into 2.4.72 gives

$$|\ell_{\omega_\alpha}(x_{\omega_\alpha}(s)) - \ell_\omega(x_\omega(s))| \leqslant K^3s^2 + M_\alpha Ks + M_\alpha$$

Proceeding by induction, we find

$$|x_{\omega_\alpha}(t) - x_\omega(t)| \leqslant 2\frac{K^nt^n}{n!} + \frac{M_\alpha}{K}\{Kt + \frac{K^2t^2}{2!} + \dots + \frac{K^{n-1}t^{n-1}}{(n-1)!}\}$$

for $0 \leqslant t \leqslant t_0$ and

$$|\ell_{\omega_\alpha}(x_{\omega_\alpha}(s)) - \ell_\omega(x_\omega(s))| \leqslant 2\frac{K^{n+1}s^n}{n!} + M_\alpha\{1 + Ks + \dots + \frac{K^{n-1}s^{n-1}}{(n-1)!}\}$$

for $0 \leqslant s \leqslant t_0$, $n = 1, 2, \dots$. Thus, in the limit $n \to \infty$,

$$(2.4.74) \quad |x_{\omega_\alpha}(t) - x_\omega(t)| \leq \frac{M_\alpha}{K}(e^{Kt} - 1) .$$

for $0 \leqslant t \leqslant t_0$. But since $\ell_{\omega_\alpha}(s) = \ell(S_s\omega_\alpha)$ converges pointwise in s to $\ell_\omega(s) = \ell(S_s\omega)$ and these functions are uniformly Lipschitz continuous by 2.4.69, the convergence is uniform on compacts. It follows from 2.4.73 that $M_\alpha \to 0$ as $\alpha \to \infty$. Thus, from 2.4.74, $x_{\omega_\alpha} \to x$ uniformly on $[0, t_0]$. This establishes 2.4.65 in Case 1.

Case 2 $\nu(\omega) = +\infty$, i.e. $\omega \in \Omega_0$

By passing to subnets it suffices to consider the following two subcases.

Case 2.1 $\nu(\omega_\alpha)$ is uniformly bounded.

It follows from 2.4.61 and 2.4.58 that the functions x_{ω_α} are uniformly bounded on compact intervals, e.g.

$$|x_{\omega_\alpha}(t_\alpha)| \leqslant K$$

for all α. Passing to subnets, we may assume that $x_{\omega_\alpha}(t_\alpha) \to x$, and it follows from the joint continuity of S that

$$T_{t_\alpha}\omega_\alpha = S_{x_{\omega_\alpha}(t_\alpha)}\omega_\alpha \to S_x\omega = \omega = T_t\omega$$

Case 2.2 $\nu(\omega_\alpha) \geqslant 1$ for all α. Then

$$T_{t_\alpha}\omega_\alpha \subseteq S_{\mathbb{R}}\,\omega_\alpha = S_{[0,\,1]}\,\omega_\alpha$$

for all α. By compactness of [0, 1], we deduce

$$T_t\omega_\alpha \to \omega$$

as in case 2.1.

This ends the proof that T is a flow. Let δ_T be the generator of this flow. We argue that $\delta \subseteq \delta_T$. By Lemma 2.4.3 it suffices to show that

$$\lim_{t\to 0}(f(T_t\omega) - f(\omega))/t = \ell(\omega)(\delta_0 f)(\omega)$$

for $f \in \mathcal{A}_\infty$, $\omega \in \Omega\backslash\Omega_0$. But

$$\begin{aligned}(f(T_t\omega) - f(\omega))/t &= (f(S_{x_\omega(t)}\omega) - f(\omega))/t \\ &\to (\delta_0 f)(\omega)x_\omega'(0) \\ &= \ell(\omega)(\delta_0 f)(\omega)\end{aligned}$$

by 2.4.58 and the chain rule.

This establishes that δ extends to the generator of a flow T.

We give a rough outline of the proof that the generator δ_T of T is equal to the closure $\overline{\delta}$ of δ. One extends δ to its natural domain, i.e. the set of $f \in D(\delta_0)$ such that the right hand side of 2.4.60 defines a function in $C_0(\Omega)$. One introduces spaces of functions which are constant on orbits of sufficiently high frequency, i.e.

(2.4.75) $\mathcal{D}_n = \{f \in \mathcal{A}_n \mid f$ has compact support and there exists an $M > 0$ such that $f(S_t\omega) = f(\omega)$ whenever $\nu(\omega) \geq M$ and $t \in \mathbb{R}\}$

Since ℓ is bounded on sets of bounded frequency by 2.4.61, it follows that $\mathcal{D}_n \subseteq D(\delta)$ for $n = 1, 2, \ldots$, and an argument based on Lemma 2.4.10 and the Stone-Weierstrass theorem shows that $\mathcal{D}_\infty$ is dense in $\mathcal{A}$. Furthermore, one can establish the formula.

(2.4.76) $\delta_0(f \circ T_t)(\omega) = \exp\{\int_0^t ds(\delta_0\ell)(T_s\omega)\}(\delta_0 f)(T_t\omega)$

for $f \in D(\delta)$, $\omega \in \Omega\backslash\Omega_0$, where the two first δ_0's are interpreted as pointwise derivatives in the sense of Theorem 2.4.20. This formula depends on differentiability of ℓ and it can be used to show that $\mathcal{D}_1$ is invariant under the automorphism group τ defined by T. But as $\mathcal{D}_1 \subseteq D(\delta_T)$, it follows from [BR], Corollary 3.1.7 that $\mathcal{D}_1$ is a core for $D(\delta_T)$, thus δ_T is the closure of δ defined on its maximal domain. But a simple regularization argument then shows that $\mathcal{D}_\infty \subseteq \mathcal{A}_\infty$ is a core for δ_T.

Remark 2.4.25

If ℓ is a continuous function on $\Omega\setminus\Omega_0$, we can define a derivation δ by

$$(\delta f)(\omega) = \begin{cases} \ell(\omega)(\delta_0 f)(\omega) & \text{for} \quad \omega \in \Omega\setminus\Omega_0 \\ 0 & \text{for} \quad \omega \in \Omega_0 \end{cases}$$

with domain $D(\delta)$ equal to the set of $f \in D(\delta_0)$ such that the right hand side defines a function in $C_0(\Omega)$. If ℓ satisfies bounds on the type

$$(2.4.75) \quad |\ell(S_t\omega) - \ell(\omega)| \leq K(\nu(\omega))|t|$$

where $K: \mathbb{R}_+ \to \mathbb{R}_+$ is a function which is bounded on bounded intervals, and for any compact subset $C \subseteq \Omega$ there exists an $\varepsilon > 0$ such that ℓ is uniformly bounded on

$$C \cap \{\omega \in \Omega | \nu(\omega) < \varepsilon\} ,$$

then it is proved in [BDGR 1], Theorem 2.6 that δ is densely defined, and there exists a unique flow T on Ω such that the generator δ_T extends δ. The significance of 2.4.75 is clear from the proof of theorem 2.4.24, while the last condition is used in the Case 2 - part of the proof, and is necessary to eliminate examples like $\Omega = \mathbb{R}^2$, $\delta_0 = y\frac{d}{dx}$, $\ell(x, y) = 1/y$, where δ is not even densely defined. If one in addition assumes that ℓ is once differentiable, it is established in [BDGR 1], Theorem 2.12 that δ_T is the closure of δ. Recently it has been established in [Rob 1] that $\bar{\delta} = \delta_T$ under weaker differentiability conditions, but stronger boundedness conditions on ℓ, i.e. it is assumed that there is a continuous function $K: \mathbb{R}_+ \to \mathbb{R}_+$ such that $|\ell(\omega)| \leq K(\nu(\omega))$ and $|\ell(S_t\omega) - \ell(\omega)| \leq K(\nu(\omega))|t|$ for all $\omega \in \Omega$. The proof is based on replacing ℓ by the regularization $\ell_\varepsilon(\omega) = \frac{1}{\varepsilon}\int_0^\varepsilon dt\,\ell(S_t\omega)$, and taking the limit $\varepsilon \to 0$. As a corollary, Theorem 2.4.24 is also valid for $\delta \in \mathrm{Der}\,(\mathcal{A}_\infty, \mathcal{A}_{1-})$ where

$$\mathcal{A}_{1-} = \{f \in C_0(\Omega) |\ \overline{\lim_{t\to 0}}\ \|(\sigma_t(f) - f)/t\| < +\infty\} .$$

Note that if δ is a derivation of the form $\delta = \ell\delta_0$, where ℓ is a real continuous function on $\Omega\setminus\Omega_0$ which is bounded, and bounded away from zero then for $\varepsilon > 0$ (resp. $\varepsilon < 0$) sufficiently small, $\delta_0 - \varepsilon\delta$ is relatively bounded by δ_0 with relative bound less than 1 on the component of Ω where $\ell > 0$ (resp. $\ell < 0$), and as δ clearly is well behaved, δ is a generator by perturbation theory, see [BR 1], Theorem 3.1.32. This argument does not depend on Lipschitz continuity of ℓ, and thus this form of continuity is only essential near the zeros of ℓ, as is indicated by the preliminary remarks to §2.4.2. Thus, using perturbation results, if ℓ is a function on $\Omega\setminus\Omega_0$ of the form $\ell = \ell_1\ell_2$ where ℓ_1 satisfies 2.4.61 and 2.4.63 and ℓ_2 is bounded and bounded away from zero, then $\delta = \ell\delta_0$ is a pregenerator.

Under suitable restrictions on the dynamics, Theorem 2.4.24 can be generalized from $G = \mathbb{R}$ to a general Lie group G.

Theorem 2.4.26

Let $A = C_0(\Omega)$ be an abelian C^*-algebra, G a Lie group, $\sigma: G \mapsto \mathrm{Aut}(A)$ a strongly continuous, locally free action of G on A with associated G-flow S, and let $\delta \in \mathrm{Der}(A_\infty, A_1)$.

It follows that δ is a pre-generator, and if T is the flow generated by δ, then $T_{\mathbb{R}}\omega \subseteq S_G\omega$ for each $\omega \in \Omega$.

Proof

The proof is based on a slight refinement of Theorem 2.4.15, and existence theorems for flows generated by Lipschitz vector-fields on manifolds in place of Lemma 2.4.22. From Theorem 2.4.15 it follows that δ has the form

$$(\delta f)(\omega) = \sum_{k=1}^{d} \ell_k(\omega)(\delta_k f)(\omega)$$

for all $f \in A_\infty$ and $\omega \in \Omega$, where $\{\delta_1, \ldots, \delta_d\}$ is a basis for the action of $\mathfrak{g}$ on A and ℓ_i are continuous functions on Ω. Using a similar argument as when the action σ is uniformly locally free, one shows that the functions ℓ_k are uniformly bounded on each G-orbit $T_G\omega$ (In the case that $T_G\omega$ is closed in Ω, it is homoemorphic to G/G_ω and exactly the same argument as in the uniformly locally free case can be used on $C_0(G/G_\omega)$, if $T_G\omega$ is not closed, more care must be taken). Since $\delta(A_\infty) \subseteq A_1$, one establishes that the functions ℓ_k are once continuously differentiable with respect to the Lie group action, and the derivatives are also uniformly bounded on each orbit. It follows from local existence theorems for flows generated by Lipschitz vector fields that for each $\omega \in \Omega$ we can construct a flow $T_t\omega$ for $|t|<\varepsilon$ in a similar manner as in Theorem 2.4.24, see [lrw 1], Theorem 3.22. But as the first under derivatives of the ℓ_k's are bounded on each orbit, the ℓ_k's are uniformly Lipschitz on each orbit, and hence the constant ε is independent of the particular ω in any given orbit. Thus a global flow T_t exists in each orbit. The remainder of the proof is as in Theorem 2.4.24.

Problem 2.4.27

Work out the details of the above argument.

In the case that the Lie group G is compact, the problems corrected with the variation of the stabilizer subgroups G_ω with ω is less severe. This is reflected in the following theorem.

Theorem 2.4.28 (Bratteli-Digernes-Goodman-Robinson)

Let $\mathcal{A} = C_0(\Omega)$ be an abelian C*-algebra, G a compact Lie group, $\sigma: G \mapsto \mathrm{Aut}(\mathcal{A})$ a strongly continuous action of G on $\mathcal{A}$ with associated flow S, and let $\delta \in \mathrm{Der}(\mathcal{A}_\infty, \mathcal{A}_1)$.

It follows that δ is a pre-generator, and if T is the flow generated by δ, then $T_{\mathbb{R}}\omega \subseteq S_G\omega$ for each $\omega \in \Omega$.

Proof

We follow the proof of Theorem 5.1 in [BDGR 1] and Theorem 3.2 in [GJ 1].

Fix a G-orbit $\mathcal{O} = S_G\omega_0$ in Ω. We first argue that $\mathcal{O}$ is a restriction set for δ, see Definition 1.6.22. $\mathcal{O}$ is compact, thus closed. If $f \in \mathcal{A}_\infty$ and $f|_{\mathcal{O}} = 0$, then f can be approximated in the $\mathcal{A}_\infty$ - topology by functions of the form fh, where $h:\Omega \mapsto [0, 1]$ is a continuous function with compact support which is constant on orbits. This can be seen as follows: As G is compact, there is for any $\varepsilon > 0$ a neighborhood V of ω_0 such that $|f(\omega)| < \varepsilon$ for $\omega \in S_G V$, and also there is a G-invariant open subset W such that $\Omega \backslash W$ is compact and $\|f|_W\| < \varepsilon$. Let $k \in C_{00}(\Omega)$ be a function such that $k = 0$ in a neighborhood of $\mathcal{O}$, $k = 1$ on $\Omega\backslash(S_G V\cup W)$ and $0 \le k \le 1$ everywhere. Put $h = \int dg\, \sigma_g(k)$. Then h is constant on G-orbits, $h \in C_{oo}(\Omega)$, $h = 1$ on $\Omega\backslash(S_G V\cup W)$ and h vanishes in a neighborhood of $\mathcal{O}$. Also $\| f\cdot h - f\| < \varepsilon$. Repeating this construction with f replaced by a finite number of f's, we may for any $\varepsilon > 0$ find a h with the above properties such that

$$\| \delta_{i_1} \cdots \delta_{i_m}(f)h - \delta_{i_1} \cdots \delta_{i_m}(f)\| < \varepsilon$$

for $m = 0, \ldots, n$, $i_j \in \{1, \ldots, d\}$, for any finite n. Here $\{\delta_1, \ldots, \delta_d\}$ is a basis for the action of $\mathfrak{g}$ on $\mathcal{A}$. But as h is σ-invariant, we have $\delta_{i_1} \cdots \delta_{i_m}(f)h = \delta_{i_1} \cdots \delta_{i_m}(fh)$, and this ends the proof that f can be approximated in the $\mathcal{A}_\infty$-topology by elements of the form fh. It now follows from Theorem 2.3.1 and Lemma 2.4.5, just as in the beginning of the proof of Theorem 2.4.11, that $\delta f|_{\mathcal{O}} = 0$. Thus $\mathcal{O}$ is a restriction set.

It follows that δ restricts to a derivation on $C(\mathcal{O})$. But $\mathcal{O}$ is canonically homeomorphic to the manifold G/G_{ω_o}, and it is clear from Lemma 2.4.10 that the set of restrictions to $\mathcal{O}$ of functions in $\mathcal{A}_\infty$ identifies with $C^\infty(G/G_{\omega_o})$. Thus $\delta|_{\mathcal{O}}$ identifies with a derivation from $C^\infty(G/G_{\omega_o})$ into $C^1(G/G_{\omega_o})$. By the multidimensional version of Lemma 2.4.22, see [lrw 1], Theorem 3.43, $\delta|_{\mathcal{O}}$ generates a flow T on $\mathcal{O}$. Thus we have constructed a one-parameter group $t \in \mathbb{R} \mapsto T_t$ of bijections T of Ω, such that the restriction of T to each G-orbit is a flow, and such that

$$(2.4.77) \quad (\delta f)(\omega) = \lim_{t\to o} (f(T_t\omega) - f(\omega))/t$$

for each $f \in \mathcal{A}_\infty$ and $\omega \in \Omega$.

Rather than explicitely patching together the flows T on the G-orbits as in the proof of Theorem 2.4.24, we can in this case use global criteria for δ to be a pre-generator. First it is clear from (2.4.77) and Lemma 2.4.10 that δ is conservative, see Proposition 2.3.4. To show that $\bar{\delta}$ is a generator, it remains to check that $(1\pm\delta)(\mathcal{A}_\infty)$ are dense in $\mathcal{A}$, see Theorem 1.5.2. To this end, let $F \in C_{oo}(\Omega)$ and let K be a G-invariant compact subset of Ω such that F=0 outside K. If $\mathcal{O}$ is a G-orbit in Ω, then there is for any $\varepsilon > 0$ an $f \in \mathcal{A}_\infty$ such that $\|(f+\delta(f))|_{\mathcal{O}} - F|_{\mathcal{O}}\| < \frac{\varepsilon}{2}$, since $\delta|_{\mathcal{O}}$ is a pre-generator. By a compactness argument, there is a G-invariant open subset $U \supseteq O$ such that

$$|f(\omega)+\delta(f)(\omega)-F(\omega)| < \varepsilon$$

for $\omega \in U$. By compactness of K, there is an open covering $\{U_1, \ldots, U_n\}$ of K of G-invariant open subsets and functions $f_i \in A_\infty$

(2.4.78) $$|f_i(\omega)+\delta(f_i)(\omega)-F(\omega)| < \varepsilon$$

for $\omega \in U_i$, $i = 1, \ldots, n$.

Using compactness of G as in the early part of the proof, we can find a partition of unity $\{h_1, \ldots, h_n\}$ subordinate to $\{U_1, \ldots, U_n\}$ such that the functions $h_1, \ldots, h_n \in C_{oo}(\Omega)$ are constant on G-orbits. Then $h_i \in A_\infty$, and we have

(2.4.79) $$\delta(h_i) = 0$$

by 2.4.77, for $i = 1, \ldots, n$. Define

$$f = \sum_{i=1}^{n} h_i f_i$$

Then $f \in A_\infty$, and

$$\delta(f) = \sum_{i=1}^{n} h_i \delta(f_i)$$

by 2.4.79. Now $F = \sum_{i=1}^{n} h_i F$, and thus

$$f+\delta(f)-F = \sum_{i=1}^{n} h_i(f_i+\delta(f_i)-F)$$

It follows from 2.4.78 that

$$\|f+\delta(f)-F\| < \varepsilon$$

and this ends the proof that $(1+\delta)(A_\infty)$ is dense in A. Similarly $(1-\delta)(A_\infty)$ is dense. Hence $\bar{\delta}$ is a generator.

Still, the flow T^δ generated by $\bar{\delta}$ has not been identified with T, so we need a separate argument showing that

$$T^\delta_{\mathbb{R}}\omega \subseteq S_G\omega$$

for each $\omega \in \Omega$. Suppose ad absurdum this is false for some $\omega \in \Omega$. In this case there is easy to construct a function $f \in C_0(\Omega)$ which is constant on G-orbits, but not constant on $T^\delta_{\mathbb{R}}\omega$. But by the first part of the proof, $\delta(f)|_{\mathcal{O}} = 0$ for each G-orbit $\mathcal{O}$, and hence $\delta(f) = 0$. In particular f is analytic for δ and

$$f(T^\delta_t\omega) = [f + \sum_{k\geq 1} \frac{t^k \delta^k(f)}{k!}](\omega)$$

$$= f(\omega),$$

a contradiction.

Once one knows $T^\delta_{\mathbb{R}}\omega \subseteq S_G\omega$, it follows from the uniqueness of the flow generated by a Lipschitz vector field on a manifold that $T^\delta = T$.

Problem 2.4.29

Let $\mathcal{A} = C_0(\Omega)$ be an abelian C*-algebra, G a Lie group, $\sigma: G \mapsto \mathrm{Aut}(\mathcal{A})$ an action of G on $\mathcal{A}$, and let $\delta \in \mathrm{Der}(\mathcal{A}_\infty, \mathcal{A}_1)$. Does it follow that δ is a pre-generator?

§2.4.3 Invariant derivations

We end this chapter with three theorems on invariant derivations on abelian C*-algebras which are not covered by the results in §2.6 and 2.8.

Theorem 2.4.30 (Kishimoto-Robinson)

Let $\mathcal{A} = C_0(\Omega)$ be an abelian C*-algebra, $\sigma: \mathbb{R} \mapsto \mathrm{Aut}(\mathcal{A})$ an action on $\mathcal{A}$ with associated flow S, and let Ω_0 be the set of fixed points for S.

Let δ be a closed derivation on $\mathcal{A}$.

The following conditions are equivalent:

1. a. δ commutes with σ, i.e. $\sigma_t(D(\delta)) = D(\delta)$ for all $t \in \mathbb{R}$ and

$$\delta\sigma_t(f) = \sigma_t\delta(f)$$

for all $f \in D(\delta)$.

b. $A_F \subseteq D(\delta)$, where A_F is the set of $\mathbb{R}$-finite elements in A (see Definition 2.2.13).

2. a. δ is a generator

b. There exists a continuous, real valued, S-invariant continuous function ℓ on $\Omega\backslash\Omega_o$ such that the flow T generated by δ on Ω is given by

$$T_t\omega = \begin{cases} S_{\ell(\omega)t}\,\omega & \text{if } \omega \;\; \Omega\backslash\Omega_o\,, \\ \omega & \text{if } \omega \;\; \Omega_o\,. \end{cases}$$

c. The function ℓ is uniformly bounded on sets of bounded frequency.

Proof

This is Theorem 2.3 in [KR 1]. Note that if 1.b. is replaced by " $A_\infty \subseteq D(\delta)$" and 2.c. by "The function ℓ is polynomially bounded in the frequency", the resulting theorem is a corollary of Theorem 2.4.24. The proof follows by combining the proof of Theorem 2.4.24 with some techniques which will be used in §2.8. We therefore omit the details.

Problem 2.4.31

Adopt the hypotheses in the first paragraph of Theorem 2.4.30. If $\delta \in \mathrm{Der}(A_F, A_1)$, is δ a pre-generator?

Theorem 2.4.32 (Goodman-Jørgensen)

Let A be an abelian C*-algebra, G a compact group, and $\sigma: G \to \mathrm{Aut}(A)$ an action of G on A. Let δ be a closed derivation of A such that

$$\sigma_g(D(\delta)) = D(\delta) \quad \text{and} \quad \sigma_g\delta = \delta\sigma_g \quad \text{for all} \quad g \in G$$

$$A^\alpha \subseteq D(\delta) \quad \text{and} \quad \delta\big|_{A^\alpha} = 0\,.$$

It follows that δ is a generator, and if $\omega \in \mathrm{Spec}\,(A)$, then the orbit of ω under the flow generated by δ is contained in the orbit of ω under the flow determined by σ.

Proof

This is Theorem 3.2 in [GJ 1]. The proof is to show that each G-orbit is a restriction set of δ as in the proof of Theorem 2.4.28, and then Theorem 2.4.33 implies that the restriction of δ to each orbit is a generator of a flow on the orbit. Finally one establishes that δ is a generator as in the last part of the proof of Theorem 2.4.28.

Theorem 2.4.33 (Goodman)

Let G be a locally compact group, and let δ be a closed derivation of $C_0(G)$ which commutes with left translations by elements in G.

It follows that δ is a generator, and there is a continuous one-parameter subgroup $\theta(t)$ of G such that

$$(e^{t\delta}f)(g) = f(g\theta(t))$$

for all $f \in C_0(G)$, $g \in G$, $t \in \mathbb{R}$.

Proof

This was proved in [Goo 2] using Lie group methods; independently it was proved for compact groups in [Nak 1] using Tannaka's duality theorem. If G is compact and abelian a very simple proof, found in [Sak 5], is as follows: Let $\gamma \in \hat{G}$, then $\gamma \in C(G)$ and $\gamma \in D(\delta)$ since $A^{\alpha}(\gamma)$ is the one-dimensional span of γ, see Lemma 2.5.8. Also, as $\delta(A^{\alpha}(\gamma)) \subseteq A^{\alpha}(\gamma)$ by Lemma 2.5.8, there is a scalar $L(\gamma) \in \mathbb{C}$ such that

$$\delta(\gamma) = L(\gamma)\gamma$$

But the derivation property

$$\delta(\gamma_1\gamma_2) = \delta(\gamma_1)\gamma_2 + \gamma_1\delta(\gamma_2)$$

implies that

$$L(\gamma_1 + \gamma_2) = L(\gamma_1) + L(\gamma_2)$$

i.e. $\gamma \to L(\gamma)$ is an additive character. As $\delta(\overline{\gamma}) = \overline{\delta(\gamma)}$ we deduce that $L(\gamma) \in \mathbb{R}$. Thus, if $t \in \mathbb{R}$, $\gamma \to e^{itL(\gamma)}$ is a character, and by Pontryagin's duality theorem there exists a $\theta(t) \in G$ such that

$$e^{itL(\gamma)} = \gamma(\theta(t))$$

The lemma is now easily verified. For the general argument, see [Goo 2].

Remark 2.4.34

A lot of work has been done on left invariant dissipative operators δ on $C_0(G)$, where G is a locally compact group. In [Rot 1], Theorem II.3.3, it is proved that each such δ extends to the generator of a strongly continuous operameter group of contractions on $C_0(G)$. In [Nak 2], it is proved that this extension is actually δ itself if G contains an open subgroup which is the direct product of an abelian group and a compact group. In [Nak 2] it is also given an example with G equal to the 3-dimensional Heisenberg group where δ is not a generator (In the example, δ is a restriction of a derivation). Thus it is essential in Theorem 2.4.33 that δ is a derivation.

§2.5 Some technical lemmas

In this section we will collect some technical lemmas which will be used later.

§2.5.1 Approximate identities

If α is an action (as a strongly continuous group of *-automorphisms) of a compact group G on a C*-algebra A, we will use the notations P_γ, $P_{ij}(\gamma)$, $A^\alpha(\gamma)$ and $A_n^\alpha(\gamma)$ introduced in Definitions 2.2.10 - 12. Then $A^\alpha \equiv A^\alpha(1)$ is the fixed point algebra under the action α. Note that $(A^\alpha)^n$ acts on $A_n^\alpha(\gamma)$ by left and right multiplication, defined by

$$(2.5.1)\qquad \begin{aligned}(x_1, \ldots, x_n)[y_{ij}] &= [x_i y_{ij}]\\ [y_{ij}](x_1, \ldots, x_n) &= [y_{ij} x_i]\end{aligned}$$

for each $(x_1, \ldots, x_n) \in (A^\alpha)^n$, $[y_{ij}] \in A_n^\alpha(\gamma)$. In particular A^α acts by left and right multiplication on $A^\alpha(\gamma)$ and on $A_1^\alpha(\gamma)$ for each $\gamma \in \hat{G}$.

If $x = (x_1, \ldots, x_{d(\gamma)})$, $y = (y_1, \ldots, y_{d(\gamma)})$ are elements in $A_1^\alpha(\gamma)$, it follows from 2.2.42 that

$$(2.5.2)\qquad xy^* \equiv \sum_{i=1}^{d(\gamma)} x_i y_i^* \in A^\alpha$$

Let $A_1^\alpha(\gamma)\, A_1^\alpha(\gamma)^*$ denote the linear subspace of A^α generated by all xy^*, with $x, y \in A_1^\alpha(\gamma)$.

Lemma 2.5.1

The space $A_1^\alpha(\gamma)\, A_1^\alpha(\gamma)^*$ is an ideal in A^α. This ideal has an approximate identity e_β of the form

$$(2.5.3)\qquad e_\beta = \sum_{i=1}^{n(\beta)} x_i^\beta x_i^{\beta *}$$

where $x_i^\beta \in A_1^\alpha(\gamma)$ for each i, β, and each $n(\beta)$ is finite.

Proof

We follow [Bro 1], [KT 1] and [BEv 1]. Since $A^{\alpha} A_1^{\alpha}(\gamma) \subseteq A_1^{\alpha}(\gamma)$, $A_1^{\alpha}(\gamma) A_1^{\alpha}(\gamma)^*$ is a two-sided ideal in A^{α}. If $x_i = (x_{i,1}, \ldots, x_{i,d})$, $i = 1, \ldots, m$ is a finite set in $A_1^{\alpha}(\gamma)$, define

$$p_x = \sum_{i=1}^{m} x_i x_i^*$$

and

$$e_\beta = e_{x,n} = np_x(\mathbb{1} + np_x)^{-1} = \sum_{i=1}^{m} y_i y_i^*$$

where $y_i = n^{\frac{1}{2}} (\mathbb{1} + np_x)^{-\frac{1}{2}} x_i \in A_1^{\alpha}(\gamma)$.

If the set of $\beta = (x, n)$, where x is a finite set in $A_1^{\alpha}(\gamma)$ and n is a positive integer, is directed by requiring

$$(x, n) < (y, k) \Leftrightarrow x \subseteq y \text{ and } n \leqslant k ,$$

then $\beta \to e_\beta$ is a net. It is straightforward from spectral theory that $\beta \to e_\beta$ is increasing and

$$\lim_{\beta\to\infty} e_\beta x = x$$

for all $x \in A_1^{\alpha}(\gamma)$, see [BR 1], Proposition 2.2.18. Thus $\beta \to e_\beta$ is an approximate identity for $A_1^{\alpha}(\gamma) A_1^{\alpha}(\gamma)^*$.

Lemma 2.5.2

Let G be a compact group, and α an action of G on a C*-algebra A. Any approximate identity for the fixed point algebra A^{α} is also an approximate identity for A, and the multiplier algebra $M(A^{\alpha})$ is contained in $M(A)$ (see Definition 1.2.1).

Proof

We follow [KT 1], Lemma 4.2 and [BEv 1], Lemma 4.1.

Let e_β be an approximate identity for A^{α}. If $x \in A_1^{\alpha}(\gamma)$, then $xx^* \in A^{\alpha}$, and hence

$$(e_\beta x - x)(e_\beta x - x)^* = e_\beta xx^* e_\beta - e_\beta xx^* - xx^* e_\beta + xx^*$$

$$\to xx^* - xx^* - xx^* + xx^* = 0$$

as $\beta \to \infty$. Thus

$$e_\beta x \to x$$

for $x \in \mathcal{A}_1^\alpha(\gamma)$, and as $\mathcal{A}_F^\alpha$ consists of the linear combinations of matrix elements in $\mathcal{A}_1^\alpha(\gamma)$, $\gamma \in \hat{G}$, it follows that

$$e_\beta x \to x$$

for all $x \in \mathcal{A}_F^\alpha$, and thus for all $x \in \mathcal{A}$, see Proposition 2.2.14. It follows that $\beta \to e_\beta$ is an approximate identity for $\mathcal{A}$.

If $y \in M(\mathcal{A}^\alpha)$ and $x \in \mathcal{A}$, then

$$yx = \lim_\beta y(e_\alpha x) = \lim_\alpha (ye_\alpha)x \in \mathcal{A},$$

and hence $M(\mathcal{A}^\alpha) \subseteq M(\mathcal{A})$.

§2.5.2 Boundedness of partially positive maps

We first state and prove a general lemma, and then give several applications.

Lemma 2.5.3 (Kishimoto)

Let $\mathcal{A}$ be a C*-algebra containing an increasing net $\mathcal{A}_\Lambda$ of closed subspaces indexed by a lattice $\{\Lambda\}$ such that $\mathcal{D} \equiv \bigcup_\Lambda \mathcal{A}_\Lambda$ is a *-subalgebra of $\mathcal{A}$, let $\{\beta\}$ be a directed set and $\beta \to \Lambda_\beta$ a mapping from $\{\beta\}$ into $\{\Lambda\}$ and assume that there exists a net $\beta \to P_\beta$: $\mathcal{A} \to \mathcal{A}_{\Lambda_\beta}$ of linear operators such that

(2.5.4) $P_\beta(\mathcal{A}_\Lambda) \subseteq \mathcal{A}_{\Lambda \cap \Lambda_\beta}$ for all Λ, β.

(2.5.5) For each β there is a constant $D_\beta > 0$ and a net $\gamma \to \sum_{i=1}^{n(\gamma)} \lambda_i^\gamma \alpha_i^\gamma$ of linear combinations of automorphisms

$\alpha_i^\gamma \in \{\alpha \in \mathrm{Aut}(\mathcal{A}) \mid \alpha(\mathcal{A}_\Lambda) = \mathcal{A}_\Lambda$ for all $\Lambda\}$ such that $\lambda_i^\gamma > 0$ for all i, , and $\sum_{i=1}^{n(\gamma)} \lambda_i^\gamma \leqslant D_\beta$ for all γ, and

$$P_\beta(x) = \lim_\gamma \sum_{i=1}^{n(\gamma)} \lambda_i^\gamma \alpha_i^\gamma(x)$$

for all $x \in \mathcal{A}$, where the limit is in the weak topology.

(2.5.6) $P_\beta(x) \to x$ for all $x \in \mathcal{D} \equiv \bigcup_\Lambda \mathcal{A}_\Lambda$.

Further, let $\sigma: \mathcal{D} \to \mathcal{A}$ be a linear map with the properties

(2.5.7) $\sigma(x^*x) \geqslant 0$ for all $x \in \mathcal{D}$

For each Λ there is a $C_\Lambda > 0$ such that

(2.5.8) $\|\sigma(x)\| \leqslant C_\Lambda \|x\|$ for all $x \in \mathcal{A}_\Lambda$.

It follows that σ is a positive map. Moreover if $\mathcal{A}$ contains an identity $\mathbb{1} \in \mathcal{D}$, then σ is bounded and $\|\sigma\| = \|\sigma(\mathbb{1})\|$.

Proof

This is essentially Remark 3 to Lemma 1.8 of [BJKR 1], but earlier a very similar result had been proved in [HR 2], Theorem 30.2.

We must show that $\sigma(y) \geqslant 0$ for all positive y. This will be achieved by approximating y with elements of the form z^2 with $z = z^* \in \mathcal{D}$. Given $1 > \varepsilon > 0$, choose $z = z^* \in \mathcal{D}$ such that $\|\sqrt{y} - z\| < \varepsilon/(2\|y\|^{\frac{1}{2}} + 1)$. Then

$$\|y - z^2\| \leqslant (\|y\|^{\frac{1}{2}} + \|z\|) \|\sqrt{y} - z\| < \varepsilon ,$$

and from 2.5.5

$$\|P_\beta(y) - P_\beta(z^2)\| < \varepsilon D_\beta .$$

Also

$$\|\sigma(P_\beta(y)) - \sigma(P_\beta(z^2))\| < \varepsilon C_{\Lambda_\beta} D_\beta$$

from 2.5.8, and thus

(2.5.9) $\sigma(P_\beta(y)) \geq \sigma(P_\beta(z^2)) - \varepsilon C_{\Lambda_\beta} D_\beta 1 \geq -\varepsilon C_{\Lambda_\beta} D_\beta 1,$

where we used that

(2.5.10) $\sigma(P_\beta(z^2)) \geq 0$

This inequality follows from 2.5.5, in fact there exists a net $\gamma \to \sum_{i=1}^{n(\gamma)} \lambda_i^\gamma \alpha_i^\gamma$ of linear combinations of automorphisms α_i^γ leaving each $\mathcal{A}_\Lambda$ invariant such that

$$P_\beta(x) = \lim_\gamma \sum_{i=1}^{n(\gamma)} \lambda_i^\gamma \alpha_i^\gamma(x)$$

for each $x \in \mathcal{D}$, hence

$$\begin{aligned} \sigma(P_\beta(z^2)) &= \sigma(\lim_\gamma \sum_i \lambda_i^\gamma \alpha_i^\gamma(z^2)) \\ &= \lim \sum_i \lambda_i^\gamma \sigma(\alpha_i^\gamma(z^2)) \\ &= \lim \sum_i \lambda_i \sigma(\alpha_i^\gamma(z)^2) \\ &\geq 0 \end{aligned}$$

where the second equality used 2.5.8 and the last inequality follows from 2.5.7 and 2.5.5. This establishes 2.5.10 and then 2.5.9. But ε was arbitrary and hence

$$\sigma(P_\beta(y)) \geq 0$$

for all Λ. But if $y \in \mathcal{A}_\Lambda$, then $P_\beta(y) \in P_{\Lambda_\beta \cap \Lambda}(\mathcal{A}) \subseteq \mathcal{A}_\Lambda$ for all β by 2.5.4. It follows from 2.5.6 and 2.5.8 that

$$\sigma(y) = \lim_\beta \sigma(P_\beta(y)) \geq 0 .$$

The last statement of the lemma follows from Russo-Dye's theorem, see [BR 1], Corollary 3.2.6.

Lemma 2.5.3 has several consequences for a locally compact group G acting as a group α of *-automorphisms on a C^*-algebra $\mathcal{A}$. The original application was to the situation

G compact abelian, $\mathcal{D} = A_F^\alpha$, [BJKR 1],

and later it was applied to

G compact, $\mathcal{D} = A_F^\alpha$, [RST 1], see Lemma 2.5.4 below,

G abelian, $\mathcal{D} = A_F^\alpha$, [KR 1], see Lemma 2.5.6 below,

G semidirect product of an abelian Lie group by a compact Lie group, $\mathcal{D} = A_\infty$, [Dav 1], this situation is slightly less clearcut than the others, and will be treated in connection with the main application of the resulting lemma, see Theorem 2.6.8.

Lemma 2.5.4

Let G be a compact group, and α an action of G on a C^*-algebra A . Let $\sigma\colon A_F^\alpha \to A$ be a linear map satisfying

(2.5.11) $\sigma(x^*x) \geq 0$ for all $x \in A_F^\alpha$

(2.5.12) For each $\gamma \in \hat{G}$ there is a $C_\gamma > 0$ such that $\|\sigma(x)\| \leq C_\gamma \|x\|$ for all $x \in A^\alpha(\gamma)$.

It follows that σ is a positive, bounded map with bound

$$\|\sigma\| \leq C_1$$

where 1 is the trivial one-dimensional representation of G.

Proof

Let

(2.5.13) $\chi_\gamma(g) = \mathrm{tr}(\gamma(g))$, $g \in G$,

be the element in the center of $L^1(G)$ corresponding to $\gamma \in \hat{G}$. By [HR 2], Theorem 28.53, there exists an approximate unit $\beta \to h_\beta$ for $L^1(G)$ such that

Each h_β is a finite linear combination

(2.5.14) $h_\beta(g) = \sum_{\gamma\in\hat{G}} \lambda_{\beta,\gamma}\, \overline{\chi_\gamma(g)}$

of characters,

(2.5.15) $h_\beta \geq 0$, and

(2.5.16) $\int_G h_\beta(g)dg \leq 1$,

for all α. Thus, putting

(2.5.17) $P_\beta(x) = \int_G dg\, h_\beta(g)\sigma_g(x)$, $x \in \mathcal{A}$,

one has from 2.2.35 and 2.2.41 that

(2.5.18) $P_\beta(\mathcal{A}) \subseteq \mathcal{A}^\alpha(\{\gamma \in \hat{G} \mid \lambda_{\alpha,\gamma} \neq 0\}) = \mathcal{A}^\alpha(\Lambda_\beta)$,

where Λ_β is a finite subset of $\hat{G}$. As h is an approximate identity for $L^1(G)$, one has

(2.5.19) $\lim_\beta P_\beta(x) = x$

for all $x \in \mathcal{A}$ where the limit is in the weak topology. Thus, if one put $\mathcal{A}_\Lambda = \mathcal{A}^\alpha(\Lambda)$ for all finite subsets $\Lambda \subseteq \hat{G}$, and $D_\beta = 1$ for all β, the properties 2.5.4, 2.5.5 and 2.5.6 of Lemma 2.5.3 follows from 2.5.15, 2.5.17, 2.5.18 and 2.5.19. It follows from 2.2.35 that

$$\|P_\gamma\| \leq d(\gamma)^2$$

for all $\gamma \in \hat{G}$, and as

$$\mathcal{A}^\alpha(\Lambda) = \dotplus_{\gamma\in\Lambda} \mathcal{A}^\alpha(\gamma)$$

for all finite subsets Λ of $\hat{G}$, it follows from 2.5.12 that the condition

$$\|\sigma(x)\| < C_\Lambda \|x\| \quad \text{for all } x \in \mathcal{A}_\Lambda$$

is fullfilled with

$$C_\Lambda = \sum_{\gamma\in\Lambda} d(\gamma)^2 C_\gamma .$$

It follows from Lemma 2.5.3 that σ is a positive map on A_F.

The boundedness of σ is now proved as follows: Let $\delta \to e_\delta$ be an approximate identity for A^α, then e_δ is an approximate identity for A by Lemma 2.5.2. If $x = x^* \in A_F$, then

$$- \|x\| e_\delta^2 \leqslant e_\delta x e_\delta \leqslant \|x\| e_\delta^2 ,$$

and by positivity of σ

$$- \|x\| \sigma(e_\delta^2) \leqslant \sigma(e_\delta x e_\delta) \leqslant \|x\| \sigma(e_\delta^2) .$$

Therefore

$$\|\sigma(e_\delta x e_\delta)\| \leqslant \|x\| \|\sigma(e_\delta^2)\| \leqslant \|x\| C_1$$

But if $x \in A^\alpha(\Lambda)$, then $e_\delta x e_\delta \in A^\alpha(\Lambda)$ for all δ, and as $\sigma \big|_{A^\alpha(\Lambda)}$ is bounded by 2.5.8, it follows that

$$\|\sigma(x)\| \leqslant C_1 \|x\|$$

for all $x = x^* \in A_F^\alpha$. Thus σ is bounded on A_F^α, and extends by continuity to a bounded map on A. This extended map is positive by continuity, since any positive element in A has the form x^*x for some $x \in A$, which can be approximated by elements from A_F^α. But then

$$\|\sigma\| \leqslant C_1$$

on A by an argument involving Russo-Dye's theorem, see [BR 3], Lemma on page 265.

Remark 2.5.5 [BJKR 1]

Positivity of σ is not a trivial consequence of 2.5.11 because positive elements in $\mathcal{A}_F$ are not necessarily linear combinations of elements of the form x^*x, with $x \in \mathcal{A}_F$. An example showing the possible problems is the following: Let $\mathcal{A} = C([0, 1])$ and $\mathcal{D}$ the *-subalgebra of polynomials. Each $f \in \mathcal{D}$ extends by analyticity to $\mathbb{C}$ and if one defines σ by $\sigma(f)(x) = f(x+2)$, $f \in \mathcal{D}$, then $\sigma: \mathcal{D} \to \mathcal{D}$ is a *-automorphism, and in particular $\sigma(f^*f) \geq 0$ for all $f \in \mathcal{D}$. But σ is not positive on $\mathcal{D}$, nor is it norm continuous. See example 1.5.5.

Lemma 2.5.6

Let G be a locally compact abelian group, and α an action of G on a C^*-algebra $\mathcal{A}$. Let $\sigma : \mathcal{A}_F^\alpha \mapsto \mathcal{A}$ be a linear map satisfying

(2.5.20) $\sigma(x^*x) \geq 0$ for all $x \in \mathcal{A}_F^\alpha$

(2.5.21) For each compact $K \subseteq \hat{G}$, there is a $C_K > 0$ such that $\|\sigma(x)\| \leq C_K \|x\|$ for all $x \in \mathcal{A}^\alpha(\gamma)$

It follows that σ is a positive map on $\mathcal{A}_F^\alpha$. Furthermore, if $\mathcal{A}$ has an identity $\mathbb{1}$, then σ is bounded and

(2.5.22) $$\|\sigma\| = \|\sigma(\mathbb{1})\|$$

Proof

By [HR 2], Theorem 33.12, $L^1(G)$ has an approximate unit h_β such that $h_\beta \geq 0$, $\int_G dg\, h_\beta(g) = 1$ and the Fourier transform $\hat{h}_\beta$ has compact support $\Lambda_\beta \subseteq \hat{G}$ for each β. But then

$$\lim_{\beta \to \infty} \int_G dg\, h_\beta(g)\alpha_g(x) = x$$

in norm for all $x \in \mathcal{A}$ by the following reasoning: if $x \in \mathcal{A}_F^\alpha$, then taking $h \in L^1(G)$ such that $\hat{h} = 1$ in a neighborhood of $\mathrm{Spec}^\alpha(x)$, then $x = \int_G dg\, h(g)\alpha_g(x)$ by 2.2.18 and 2.2.19, and hence

$$\int_G dg\, h_\beta(g)\alpha_g(x) = \int_G dg\, (h_\beta * h)(g)\alpha_g(x) \to \int_G dg\, h(g)\alpha_g(x) = x$$

since h_α is an approximate identity for $L^1(G)$. But as $\mathcal{A}_F^\alpha$ is norm-dense in A, the claim follows.

The rest of the proof is now identical with the proof of Lemma 2.5.4.

Problem 2.5.7

Does boundedness of σ follow in Lemma 2.5.6, even when A does not have an identity?

The preceding lemmas have immediate applications to closed derivations commuting with group actions. For these applications we need the following.

Lemma 2.5.8

Let G be a locally compact group, and α a strongly continuous action of G as a group of isometries on a Banach space A. Let δ be a closed operator on A which commutes with the action α in the sense that

$$\begin{aligned} &\alpha_g(D(\delta)) = D(\delta)\,, \\ &\delta\alpha_g(x) = \alpha_g\delta(x) \end{aligned} \tag{2.5.23}$$

for all $x \in D(\delta)$, $g \in G$. Then, if $f \in L^1(G)$

$$\begin{aligned} &\alpha_f(D(\delta)) \subseteq D(\delta)\,, \\ &\delta\alpha_f(x) = \alpha_f\delta(x) \end{aligned} \tag{2.5.24}$$

for all $x \in D(\delta)$, where α_f is defined by 1.5.6.

If G is compact, then

$$\delta(A^\alpha(\gamma) \cap D(\delta)) \subseteq A^\alpha(\gamma) \quad \text{for all} \quad \gamma \in \hat{G}\,. \tag{2.5.25}$$

If G is abelian, then

$$\delta(A^\alpha(K) \cap D(\delta)) \subseteq A^\alpha(K) \quad \text{for all closed subsets} \quad K \subseteq \hat{G}\,. \tag{2.5.26}$$

If G is abelian or compact, then

(2.5.27) $\quad A_F^{\alpha} \cap D(\delta)$ is a core for δ .

If G is a Lie group, then

(2.5.28) $\quad \{x \in A_\infty \cap D(\delta) \mid \delta(x) \in A_\infty\}$ is a core for δ .

Remark 2.5.9

Note that under the assumptions in the lemma, it does not follow in general that

$$\delta(A_\infty \cap D(\delta)) \subseteq A_1$$

when G is a Lie group. An example is the flow an $\mathbb{T} \times \mathbb{R}$ defined in Remark 2.4.16, with generator $\delta_0 = -i\, x \frac{\partial}{\partial\theta}$. Let δ be the derivation $\delta = -ie^{x^2} \frac{\partial}{\partial\theta}$ with its natural domain, and let $f(e^{i\theta}, x) = \frac{e^{-x^2}}{1+|x|} e^{i\theta}$.

Then δ commutes with the flow generated by δ_0 , $f \in A_\infty$, $f \in D(\delta)$, but $(\delta f)(e^{i\theta},x) = \frac{1}{1+|x|} e^{i\theta}$, so $\delta f \notin D(\delta_0) = A_1$

Proof of Lemma 2.5.8.

The relation 2.5.24 is immediate from closability of δ and the definition of α_f , see [BR 1], Propositions 2.5.18 and 3.1.4. Then 2.5.24 together with 2.2.22 and 2.2.11 implies 2.5.26; and 2.5.24 together with 2.2.35 and 2.2.41 implies 2.5.25. The result 2.5.27 follows by fixing an $x \in D(\delta)$ and then approximate x by $\alpha_f(x)$, where f runs through an approximate unit for $L^1(G)$ such that $\hat{f}$ has compact support in the abelian case, and f is a trigonometric polynomial in the compact case, see [HR 2], Theorems 33.12 and 28.53. Finally, when G is a Lie group we use the same procedure with an approximate unit $f \in C_{00}^\infty(G)$.

Definition 2.5.9

Let G be a group, and α an action of G as isometries on a Banach space A. We say that an operator $\delta : D(\delta) \subseteq A \mapsto A$ commutes with α , and write $[\delta,\alpha] = 0$, if

$$\begin{aligned} &\alpha_g(D(\delta)) = D(\delta) \\ (2.5.29) \qquad & \\ &\delta\alpha_g(x) = \alpha_g\delta(x) \end{aligned}$$

for all $x \in D(\delta)$, $g \in G$.

Note that if δ is closable and $[\delta, \alpha] = 0$, then $[\bar{\delta}, \alpha] = 0$, where $\bar{\delta}$ is the closure of δ . It follows from Lemma 2.5.8 that if δ is a closed, densely defined operator commuting with an action α of a locally compact abelian or compact group G , then $\delta|_{A^\alpha(K)}$ is a densely defined closed operator from $A^\alpha(K)$ into $A^\alpha(K)$, for each compact $K \subseteq G$. Here we use the convention

$$(2.5.30) \qquad A^\alpha(K) = \sum_{\gamma \in K} A^\alpha(\gamma)$$

when G is compact and K is a compact (i.e. finite) subset of $\hat{G}$.

<u>Proposition 2.5.10</u> (Kishimoto-Robinson)

Let A be a C*-algebra, α an action of a locally compact group G which is abelian or compact on A, and δ a closed derivation commuting with α . Assume that the restrictions $\delta|_{A^\alpha(K)}$ generates a one-parameter group of bounded maps on $A^\alpha(K)$ for each compact $K \subseteq \hat{G}$. It follows that δ is a generator.

<u>Proof</u>

This result was first proved for compact abelian groups in [KR 2], Proposition 2, using a completely different method from the following, and the present proof of the result for abelian groups was indicated in the introduction to [KR 1].

We may assume that A has an identity by adjoining one and extending α and δ . For each compact $K \subsetneq \hat{G}$, let $t \in \mathbb{R} \mapsto \beta_t^K$ be the strongly continuous group of bounded maps of $A^\alpha(K)$ generated by $\delta^K = \delta|_{A^\alpha(K)}$. It follows from the formula 1.5.2, $e^{t\delta^K}(x) = \lim_{n\to\infty} (1-\frac{t}{n}\delta)^{-n}(x)$ for $x \in A^\alpha(K)$, that K_1 K_2 implies $\beta^{K_1} = \beta^{K_2} \mid A^\alpha(K_1)$, and thus the β^{K}'s defines a one-parameter group of linear maps β of A_F^α . Reasoning with analytic elements of the form 1.5.4. for the β^{K}'s, and using the derivation property of δ, are verifies that

$$\beta_t(xy) = \beta_t(x)\beta_t(y)$$

for $x, y \in A_F^\alpha$, while trivially $\beta_t(x^*) = \beta_t(x)$, thus β is a one-parameter group of *-automorphisms of A_F^α. In particular

$$\beta_t(x^*x) = \beta_t(x)^*\beta_t(x) \geq 0$$

It now follows, from Lemma 2.5.4 when G is compact and from Lemma 2.5.6 when G is abelian, that β is a one-parameter group of isometries. Thus β extends by continuity to a one-parameter group of *-automorphisms of A. But as $A_F^\alpha \cap D(\delta)$ is a core for δ by Lemma 2.5.8, it follows that δ is the generator of the extended β.

§2.5.3. A generalized derivation theorem.

Theorem 2.5.11 (Christensen - Evans)

Let A be a C*-algebra on a Hilbert space H, π a representation of A on another Hilbert space K, and Δ a linear map of A into $L(H,K)$ ($\equiv$ the set of bounded linear operators from H into K) such that

(2.5.31) $\Delta(x)^*\Delta(y) \in \bar{A}$ ($\equiv$ the σ-weak closure of A on H), and

(2.5.32) $\Delta(xy) = \pi(x)\Delta(y)+\Delta(x)y$

for all $x, y \in A$.

Then there exists an operator h in the σ-weak closure of the linear span of the set $\{\Delta(x)y \mid x,y \in A\}$ such that

$$\Delta(x) = hx-\pi(x)h$$

for all $x \in A$.

Remark 2.5.12

In the case $H = K$, π = identity-representation, this theorem reduces to Sakai-Kadison's derivation theorem alluded to before Theorem

1.2.3. In the case $H = K$, π = automorphism of A, it gives an implementation theorem for twisted derivations. This theorem is also instrumental for the classification of bounded complete dissipations, see §3.1.

Proof of Theorem 2.5.11

This is theorem 2.1 in [CE 1], and we follow the proof there. We may assume that $\bar{A}$ contains the identity operator $\mathbb{1}$ on H, because if the identity q of $\bar{A}$ is a projection less than $\mathbb{1}$, we remark that the linear span of $\{\Delta(x)y \mid x,y \in A\}$ identifies with a subspace of $L(qH,K)$. Consider the representation $id \oplus \pi$ of A on $H \oplus K$ and the derivation δ from $(id \oplus \pi)(A)$ into $L(H \oplus K)$ defined by

(2.5.34) $$\delta\begin{pmatrix} a & 0 \\ 0 & \pi(a) \end{pmatrix} = \begin{pmatrix} 0 & 0 \\ \Delta(a) & 0 \end{pmatrix}, \quad a \in A,$$

and define

(2.5.35) $\hat{A} = (id \oplus \pi)(A)$.

(2.5.36) $p = \begin{pmatrix} 1 & 0 \\ 0 & 0 \end{pmatrix}$ = projection from $H \oplus K$ onto H .

(2.5.37) B = C*-algebra generated by $\hat{A}$, p and $\delta(\hat{A})$.

(2.5.38) M = σ-weak closure of $\hat{A}$.

(2.5.39) N = σ-weak closure of B .

(2.5.40) R = σ-weakly closed linear span of $\{\Delta(x)y \mid x,y \in A\}$.

(2.5.41) $J = \left\{\begin{pmatrix} 0 & 0 \\ h & 0 \end{pmatrix} \mid h \in R\right\}$.

The theorem is proved via two lemmas:

Lemma 2.5.13

$$J = (1-p)Np .$$

Lemma 2.5.14

There is an $y \in (1-p)Np$ such that

(2.5.42) $\delta(x) = yx - xy$

for all $x \in \hat{A}$..

It follows from these two lemmas that there is a $h \in R$ such that

$$\delta\begin{pmatrix} x & 0 \\ 0 & \pi(x) \end{pmatrix} = \left[\begin{pmatrix} 0 & 0 \\ h & 0 \end{pmatrix}, \begin{pmatrix} x & 0 \\ 0 & \pi(x) \end{pmatrix}\right]$$

for all $x \in A$, i.e.

$$\begin{pmatrix} 0 & 0 \\ \Delta(x) & 0 \end{pmatrix} = \begin{pmatrix} 0 & 0 \\ hx-\pi(x)h & 0 \end{pmatrix}$$

and this establishes Theorem 2.5.11.

<u>Proof</u> of Lemma 2.5.13

Define

(2.5.43) $\mathcal{D}$ = ultraweakly closed *-algebra on K generated by $\pi(A)$ and $\{\Delta(x)y\Delta(z) \mid x, y, z \in A\}$

(2.5.44) $C = \left\{ \begin{pmatrix} a & t^* \\ s & d \end{pmatrix} \middle| a \in \bar{A}, s, t \in R, d \in \mathcal{D} \right\}$

Then clearly $(1-p)Cp = J$, so if we can show that $C = N$ we are finished. But C is a σ-weakly closed self-adjoint linear subspace of N which contains the generators

$$\begin{pmatrix} x & 0 \\ 0 & \pi(x) \end{pmatrix}, \quad \begin{pmatrix} 0 & 0 \\ \Delta(y) & 0 \end{pmatrix}, \quad p = \begin{pmatrix} 1 & 0 \\ 0 & 0 \end{pmatrix}$$

of N, so it suffices to show that C is an algebra. But this follows from the assumptions 2.5.31 and 2.5.32 together with straightforward algebraic considerations. (The assumption 2.5.32 is used in the form:

$\pi(x)\Delta(y) = \Delta(xy) - \Delta(x)y \subseteq R$ for $x, y \in A$).

The proof of Lemma 2.5.14 depends on three general lemmas.

<u>Lemma 2.5.15</u> (Ringrose)

Let A be a C^*-algebra on a Hilbert space H and let $\delta : A \mapsto L(H)$ be an everywhere defined derivation. It follows that δ is continuous in the σ-weak – σ-weak topology, and in particular δ extends to a derivation of the σ-weak closure $\bar{A}$ of A into $L(H)$.

Proof

We follow [Rin 1]. First note that δ is norm continuous by a simplified version of the argument used to prove Theorem 2.3.1, see [Rin 1], Theorem 2. If $M = \bar{A}$, we have to show that the linear functional

$$x \mapsto \eta(x) = \omega(\delta(x))$$

is σ-weakly continuous for each $\omega \in M_*$ = the predual of M. But if A_1^+ is the positive part of the unit ball of A, it suffices to show that $\eta|_{A_1^+}$ is σ-strongly continuous at 0, see [Dix 1], Corollaire, p.45. But for $x \in A_1^+$,

$$\begin{aligned}\eta(x) = \omega(\delta(x)) &= \omega(\delta((x^{\frac{1}{2}})^2)\\ &= \omega(x^{\frac{1}{2}}\delta(x^{\frac{1}{2}}) + \delta(x^{\frac{1}{2}})x^{\frac{1}{2}})\end{aligned}$$

and hence

$$\begin{aligned}|\eta(x)| &\leq 2\omega(x)^{\frac{1}{2}}\omega(\delta(x^{\frac{1}{2}})^2)^{\frac{1}{2}}\\ &\leq 2\|\delta\|\omega(x)^{\frac{1}{2}}\end{aligned}$$

by Schwarz's inequality. From this the desired continuity is obvious.

Lemma 2.5.16 (Christensen)

Let M be a finite von Neumann algebra on a Hilbert space H. If $\delta : M \mapsto L(H)$ is an everywhere defined derivation, then there is a h in the σ-weakly closed convex hull of $\{\delta(u)u^* | u$ unitary in $M\}$ such that

$$\delta(x) = hx - xh$$

for all $x \in M$.

Proof

Let $U(M)$ be the unitary group in M, let C be the σ-weakly closed convex hull of the set $\{\delta(u)u^* | u \in \mathcal{U}(M)\}$, and define a representation T of $\mathcal{U}(M)$ into the affine transformations of C by

(2.5.45) $T_u(x) = u\,x\,u^* + \delta(u)u^*$

Then

$$T_u(x_1) - T_u(x_2) = u(x_1 - x_2)u^*$$

and thus for any finite trace τ on M

$$\tau((T_u(x_1)-T_u(x_2))^*(T_u(x_1)-T_u(x_2)))$$

$$= \tau((x_1-x_2)^*(x_1-x_2)) ,$$

and as the finite traces on M separates any positive element from zero , it follows that the group T is non-contractive with respect to the σ-strong topology. The Ryll-Nardzewski fixed point theorem, [Gre 1], Appendix 2, implies that the T_u have a common fixed point h in C , i.e.

$$h = uhu^* + \delta(u)u^*$$

for all $u \in \mathcal{U}(M)$. Then

$$\delta(x) = hx - xh$$

for all $x \in \mathcal{U}(M)$, and thus for all $x \in M$.

<u>Lemma 2.5.17</u> (Christensen - Evans)

Let M be a properly infinite von Neumann algebra on a Hilbert space H and $p \in M'$ a projection. If $\delta : M \mapsto L(H)$ is an everywhere defined derivation such that

(2.5.46) $\delta(M) \subseteq (\mathbb{1}-p)L(H)p$

and such that the von Neumann algebra N generated by M, $\delta(M)$ and p has the property

(2.5.47) $pNp = Mp$

then there is a $y \in (1-p)Np$ such that

$$\delta(x) = yx - xy$$

for all $x \in M$.

Proof

By a rather long argument, [Chr 1] Theorem 3.2, one can show that δ is implemented by an operator $y \in L(H)$. By a more straightforward compactness argument, [Kad 3], Lemma 4, the operator y can be chosen with totally minimal norm among the operators implementing δ . This means that for any projection e in the centre of N , ye has minimal norm among the operators implementing the derivation $x \mapsto \delta(x)e$. Since $p \in M'$

$$\delta(x) = (1-p)[y, x]p = [(1-p)yp, x]$$

for all $x \in M$, and as $p \in N$

$$\|(1-p)ype\| = \|(1-p)yep\| \leq \|ye\|$$

for all projections $e \in N \cap N'$. Thus we may and will assume

$$y = (1-p)yp .$$

We next show that

$$\|eye\| = \|yc(e)\|$$

for all projections $e \in N'$ where $c(e)$ is the central support of e in N' .

To see this let $\{e_j\}$ be a family of pairwise orthoganal projections in N' such that each e_j is dominated by e in N' in Murray-von Neumann sense, and

$$\sum_j e_j = c(e)$$

Let $v_j \in N'$ be partial isometries with $v_j v_j^* = e_j$, $v_j^* v_j \leq e$, then the operator $y_o = \sum_j v_j y v_j^*$ satisfies $\|y_o\| \leq \|eye\|$ and

$$[y_o , x] = \sum_j v_j [y, x] v_j^*$$

$$= \sum_j v_j \delta(x) v_j^*$$

$$= \delta(x)c(e) = [yc(e), x]$$

Thus by total minimality of the norm of y ,

$$\|yc(e)\| \leq \|y_o\|$$

but

$$\|y_o\| \leq \|eye\| \leq \|ye\| \leq \|yc(e)e\| \leq \|yc(e)\|$$

and we get

(2.5.48) $\quad \|ye\| = \|eye\| = \|yc(e)\|$

for all projections $e \in N'$.

It follows from the Jacobi identity

$$[[y, z], x] + [[z, x], y] + [[x, y], z] = 0 ,$$

applied to $x \in M$, $z \in N'$, that $\mathrm{ad}(y)$ maps N' into M' . Assume ad absurdum that $y \notin N$, then there is a projection $e \in N'$ such that $[y, e] \neq 0$. But as $\mathrm{ad}(y)(N') \subseteq M'$, we have $[y, e] \in M'$, and we can assume

$$0 \neq (1-e)ye = (1-e[y, e] \in M' ,$$

eventually replacing e by $1-e$.

The operator $ey^*(1-e)ye$ is then non-zero and belongs to $pM'p$ because $p \in N$, $e \in N'$ and $yp = y$. Hence

$$ey^*(1-e)ye \in pM'p = (Mp)' = (Np)' = pN'p$$

where we used 2.5.47. Thus there exists exactly one (positive) operator

$z \in N'$ such that

$$ey^*(1-e)ye = zp$$

and $zc(p) = z$. Choose a positive real λ and a projection $q \in N'$ such that $z \geq \lambda q$, then $e \geq q$ since $e \in N'$, $q \leq c(p)$ and $ep \geq qp$. The contradiction to $(1-e)ye \neq 0$ now follows from

$$\begin{aligned} \|qy^*eyq\| &= \|eyq\|^2 \geq \|qyq\|^2 \\ &= \|yq\|^2 \\ &= \|qy^*eyq + qy^*(1-e)yq\| \\ &= \|pqy^*eyqp + pqzqp\| \\ &\gneq \|pqy^*eyqp\| = \|qy^*eyq\| \end{aligned}$$

where the third step used 2.5.48. Thus $y \in N$ and Lemma 2.5.17 follow.

Proof of Lemma 2.5.14.

Note first that if we can find a $h \in N$ such that $\delta = \mathrm{ad}(h)$, then it follows from the particular form 2.5.34 of δ that $\delta = \mathrm{ad}((1-p)yp)$, hence it suffices to find a $y \in N$ with $\delta = \mathrm{ad}(y)$. By Lemma 2.5.15, δ extends by σ-weak continuity to a derivation from M into N . By Lemma 2.5.16, there is an $y \in N$ such that $y = (1-p)yp$ and the restriction of δ to the centre of M is implemented by y . Thus, perturbing δ by $-\mathrm{ad}(y)$, we may assume $\delta|\mathrm{centre}(M) = 0$. If e is the central projection corresponding to the finite part of M , it follows that δ maps Me into eNe and $M(1-e)$ into $(1-e)N(1-e)$. Since $p \in M'$, we have $pe = ep$. Since $N = C$, where C is given by 2.5.44, we have $Mp = pNp = \overline{A}$. and it follows that

$$M(1-e)p = p(1-e)N(1-e)p$$

It now follows from Lemma 2.5.16 applied to $\delta|M_e$ and Lemma 2.5.17 applied to $\delta|M_{1-e}$, that δ is implemented by an element in N . This ends the proof of Lemma 2.5.14 and thence of Theorem 2.5.11.

§2.5.4 Bounded perturbations of generator derivations

Let δ be the generator of a one-parameter group $e^{t\delta}$ of *-automorphisms on a unital C*-algebra A. If h is a skew-adjoint element in A, define

$$(2.5.49)\quad \delta^h(x) = \delta(x) + hx - xh = \delta(x) + \delta_h(x)$$

for all $x \in D(\delta)$. Then δ^h is the generator of a one parameter group of *-automorphisms, see [BR 1], Theorem 3.1.32. From [BR 2], Proposition 5.4.1, it follows that there exists a map $t \to \Gamma_t^h$ from $\mathbb{R}$ into the unitary group of A such that

$$(2.5.50)\quad e^{t\delta^h}(x) = \Gamma_t^h\, e^{t\delta}(x)(\Gamma_t^h)^*$$

for all $x \in A$, $t \in \mathbb{R}$. The map can be taken to be a 1-cocycle, i.e.

$$(2.5.51)\quad \Gamma_{t+s}^h = \Gamma_t^h\, e^{t\delta}(\Gamma_s^h)\ .$$

Here, Γ^h can be taken to be the unique solution of the differential equation

$$(2.5.52)\quad \frac{d}{dt}(\Gamma_t^h) = \Gamma_t^h\, e^{t\delta}(h)\ ,$$

with initial condition $\Gamma_0^h = \mathbb{1}$. Then Γ^h also satisfies the equation

$$(2.5.53)\quad \frac{d}{dt}(\Gamma_t^h) = e^{t\delta^h}(h)\Gamma_t^h\ .$$

Solving these equations by iteration, we find that Γ^h is given by the perturbation expansions

$$(2.5.54)\quad \Gamma_t^h = \mathbb{1} + \sum_{n\geq 1}\int_0^t dt_1 \int_0^{t_1} dt_2 \ldots \int_0^{t_{n-1}} dt_n\, e^{t_n\delta}(h) \ldots e^{t_1\delta}(h)$$

$$= \mathbb{1} + \sum_{n\geq 1}\int_0^t dt_1 \int_0^{t_1} dt_2 \ldots \int_0^{t_{n-1}} dt_n\, e^{t_1\delta^h}(h) \ldots e^{t_n\delta^h}(h)$$

when $t \geq 0$.

§2.5.4.1 Automatic smoothness of 1-cocycles

Each of the terms of the expansions 2.5.50 involves a smoothing operation on h. This smoothing suffices to ensure that Γ_t^P is always contained in the domain $D(\delta)$ of the generator δ, even when h is not contained in this domain. More precisely:

Lemma 2.5.18 (Bratteli-Jørgensen)

Adopt the assumptions and notation in the introduction to Section 2.5.4. It follows that $\Gamma_t^h \in D(\delta) = D(\delta^h)$, and

$$(2.5.55)\quad \begin{aligned} \delta(\Gamma_t^h) &= (e^{t\delta^h}(h) - h)\Gamma_t^h \\ \delta^h(\Gamma_t^h) &= \Gamma_t^h(e^{t\delta}(h) - h) . \end{aligned}$$

Hence one has

$$(2.5.56)\quad \frac{d}{dt}(\Gamma_t^h) = \delta(\Gamma_t^h) + h\Gamma_t^h = \delta^h(\Gamma_t^h) + \Gamma_t^h h$$

Proof

This is Lemma 2.1 in [BJ 2], and we follow the proof there.

First note that A may be faithfully represented on a Hilbert space H such that $e^{t\delta}$ is covariant, i.e. there exists a strongly continuous unitary group $t \to e^{tH}$ such that $e^{t\delta}(x) = e^{tH}\, x\, e^{-tH}$ for all $x \in A$. One way to prove this is to use Theorem A1 from [KR 2], which states that if f is an almost everywhere positive Lebesque-integrable function on $\mathbb{R}$ with total integral one, and ω is an arbitrary state on A , then the state

$$\omega_f = \int dt\, f(t)\, \omega \circ e^{t\delta}$$

generates a covariant representation. Since the states ω_f separates any positive element in A from zero, A has a faithful covariant representation.

Here H is the skew-adjoint generator of e^{tH}, and the cocycle Γ_t^h is then given by

(2.5.57) $\Gamma_t^h = e^{t(H+h)}e^{-tH}$,

see [BR 2], Corollary 5.4.2.

Assume first that H is bounded, and then $\delta = \mathrm{ad}(H)$ extends to all of $L(H)$. One has

(2.5.58) $$\delta(\Gamma_t^h) = \delta(e^{t(H+h)})e^{-tH} = \int_0^t ds\, e^{s(H+h)}\delta(h)e^{(t-s)(H+h)}e^{-tH}$$

by 1.6.12. Thus

$$\delta(\Gamma_t^h) = \int_0^t ds\, e^{s(H+h)}\delta(h)e^{-s(H+h)}e^{t(H+h)}e^{-tH}$$

(2.5.59) $$= \int_0^t ds\, e^{s\delta^h}(\delta(h))\Gamma_t^h = \int_0^t ds\, e^{s\delta^h}(\delta^h(h))\Gamma_t^h$$

$$= (e^{t\delta^h}(h) - h)\Gamma_t^h$$

Assume next that H is unbounded. Using spectral theory, we may find a sequence of bounded, skew-adjoint operators H_n such that $H_n\psi \to H\psi$ for all $\psi \in D(H)$. It follows from Theorem 1.5.8. that $e^{tH_n}\psi \to e^{tH}\psi$ and $e^{t(H_n+h)}\psi \to e^{t(H+h)}\psi$ for all $\psi \in H$. Hence

$$(e^{t(H_n+h)}he^{-t(H_n+h)} - h)e^{t(H_n+h)}e^{-tH_n}$$

converges strongly to

$$(e^{t\delta^h}(h) - h)\Gamma_t^h$$

But then 2.5.59 implies that $[H_n, e^{t(H_n+h)}e^{-tH_n}]$ converges strongly to $(e^{t\delta^h}(h) - h)\Gamma_t^h$, but at the same time it is clear that this expression converges as a bilinear form on $D(H) \times D(H)$ to $[H, \Gamma_t^h]$. It follows that $\Gamma_t^h \in D(\delta)$ and

$$\delta(\Gamma_t^h) = (e^{t\delta^h}(h) - h)\Gamma_t^h ,$$

see [BR 1], Proposition 3.2.55. This is the first formula of Lemma 2.5.18, and the remainding assertions of the lemma follows from the formuli

$$\frac{d}{dt}(\Gamma_t^h) = e^{t\delta^h}(h)\Gamma_t^h = \Gamma_t^h e^{t\delta}(h)$$

and $\delta^h = \delta + \delta_h$.

§2.5.4.2. Analytic elements for perturbed groups

Lemma 2.5.19 (Goodman-Jørgensen)

Let δ be a derivation of a C*-algebra or von Neumann algebra A . If $h = -h^* \in A$ define δ^h on $D(\delta)$ by 2.5.45:

$$\delta^h(x) = \delta(x) + hx - xh .$$

If x and h are analytic for δ, it follows that x is analytic for δ^h . Moreover, if t_0 is a positive number such that

$$\sum_{n=0}^{\infty} \frac{t_0{}^n \, \|\delta^n(x)\|}{n!} < +\infty \ , \qquad \sum_{n=0}^{\infty} \frac{t_0{}^n \, \|\delta^n(h)\|}{n!} < +\infty \ ,$$

then

$$(2.5.60) \qquad \sum_{n=0}^{\infty} \frac{t^n \, \|(\delta^h)^n(x)\|}{n!} < \infty$$

for all positive $t < t_0$.

Proof

This is Lemma 3.6 in [BG 1], and we follow the proof there.

A direct proof of 2.5.56 leads to seemingly insurmountable combinatorial difficulties, so we will give a proof based on the cocycle formalism for perturbations of δ. This argument does not presuppose that δ is a generator.

First define

$$e^{t\delta}(x) = \sum_{n=0}^{\infty} \frac{t^n \delta^n(x)}{n!}$$

for $|t| < t_0$, and define $e^{t\delta}(h)$ analogously. Next define

$$\Gamma_t^h = \mathbb{1} + \sum_{n\geq 1} \int_0^t dt_1 \int_0^t dt_2 \ldots \int_0^{t_{n-1}} dt_n \, e^{t_n\delta}(h) \ldots e^{t_1\delta}(h)$$

for $|t| < t_0$. Then Γ_t^h is the unique solution of the differential equation

$$\frac{d}{dt}(\Gamma_t^h) = \Gamma_t^h \, e^{t\delta}(h) \, ,$$

and the functions $t \to e^{t\delta}(h)$ and $t \to \Gamma_t^h$ clearly have analytic extensions to the disk $\{z: \ |z| < t_0\}$.

Next define

$$e^{t\delta^h}(x) = \Gamma_t^h \, e^{t\delta}(x)\Gamma_t^{h*}$$

for $|t| < t_0$. Then $t \to e^{t\delta^h}(x)$ has an analytic extension to the disk $|z| < t_0$, and we have

$$\frac{d}{dt}(e^{t\delta^h}(x)) = \Gamma_t^h e^{t\delta}(\delta(x) + hx - xh)\Gamma_t^{h*} = \Gamma_t^h e^{t\delta}(\delta^h(x))\Gamma_t^{h*} \, .$$

Iterating this computation, we get

$$\frac{d^n}{dt^n}(e^{t\delta^h}(x))\Big|_{t=0} = (\delta^h)^n(x)$$

Therefore the analytic function $e^{z\delta^h}(x)$ has Taylor series

$$e^{z\delta^h}(x) = \sum_{n\geq 0} \frac{z^n(\delta^h)^n(x)}{n!}$$

where the power series converges in norm for $|z| < t_0$, see [BR 1], Theorem 2.5.21. The Lemma follows.

§2.6 Invariant non-commutative vectorfields

As stated in §2.1, our main aim is to classify derivations mapping one class of smooth elements for a group action into another, this will be done in §2.10. A subsidiary aim is to classify closed derivations commuting with the group action, without any assumptions on the domain of the derivations, this will be done in §2.8. We saw in Lemma 2.5.8 that the last situation is almost contained in the first, there are however some results, notably for non-compact groups, which depend on δ satisfying both requirements: δ is defined on a class of smooth elements and δ commutes with the group action. We have already encountered one such result in Theorem 2.4.30, and we collect some others in this chapter.

2.6.1 Compact groups

Theorem 2.6.1 (Peligrad)

Let A be a C*-algebra, G a compact group, α an action of G on A and δ a closed derivation such that

$$[\delta, \alpha] = 0, \quad \text{(See Definition 2.5.9)}$$

and

$$A_F^\alpha \subseteq D(\delta). \quad \text{(See Definition 2.2.15)}$$

It follows that δ is a generator.

If G is abelian and the assumptions that δ is closed and $A_F^\alpha \subseteq D(\delta)$ is replaced by the single assumption that $D(\delta) = A_F^\alpha$, it follows that δ is a pregenerator.

Proof

This was originally proved in [BJ 1], Theorem 4.4 and [Pel 2], Theorem 2.5, using different methods from here.

By Lemma 2.5.8, we have that

$$\delta(A^\alpha(\gamma)) \subseteq A^\alpha(\gamma)$$

for all $\gamma \in \hat{G}$. As $A^\alpha(\gamma)$ is a closed subspace of A and δ is closed, it follows from the closed graph theorem that $\delta|_{A^\alpha(\gamma)}$ is bounded, and thus $A^\alpha(\gamma)$ consists of analytic vectors for δ. Thus $\delta|_{A^\alpha(\gamma)}$ is bounded, and hence $A^\alpha(\gamma)$ consists of analytic vectors for δ. Thus $\delta|_{A^\alpha(\gamma)}$ generates a one-parameter group of bounded maps on $A^\alpha(\gamma)$. The first part of the theorem now follows from Proposition 2.5.10.

Assume next that G is abelian, and assume only that $[\delta, \alpha] = 0$ and $D(\delta) = A_F^\alpha$. Then $\delta(A^\alpha) \subseteq A^\alpha$, thus $\delta|_{A^\alpha}$ is bounded by Corollary 1.4.10, and thus $\delta|_{A^\alpha}$ generates a norm-continuous one-parameter group of $*$-automorphisms on A^α. This group is automatically a group of isometries, thus $\delta|_{A^\alpha}$ is conservative. It follows from Lemma 2.8.14 that each of the operators $\pm\delta$ are dissipative on $A^\alpha(\gamma)$ for each $\gamma \in \hat{G}$. Thus $\delta|_{A^\alpha(\gamma)}$ is bounded for each γ by Proposition 1.4.7, and hence each $A^\alpha(\gamma)$ consists of entire analytic elements for δ. The remainder of the proof is as before.

§2.6.2 Abelian locally compact groups

Theorem 2.6.2 (Kishimoto-Robinson)

Let A be a C*-algebra, G a locally compact abelian group, α an action of G on A and δ a closed derivation such that

$$[\delta, \alpha] = 0$$

and

$$A_F^\alpha \subseteq D(\delta) . \qquad \text{(See Definition 2.2.13)}$$

It follows that δ is a generator.

Proof

This was remarked in the introduction of [KR 1]. The proof follows from Lemma 2.5.8 and Proposition 2.5.10 as in the proof of Theorem 2.6.1.

Under some general conditions on the dynamics α it is possible to classify the derivations δ occuring in Theorem 2.6.2. To this end we need some definitions.

Definition 2.6.3

Let A be a C*-algebra. A C*-subalgebra $B \subseteq A$ is said to be a hereditary subalgebra if for all $x \in A$, $y \in B$ with $0 \leq x \leq y$ we have $x \in B$. (Note in particular that if $p \in A$ is a projection, then pAp is a hereditary subalgebra of A, and if

A is a von Neumann algebra this is the most general σ-weakly closed hereditary subalgebra. If A is abelian, a closed subalgebra B is hereditary if and only if B is an ideal.) If G is a locally compact abelian group and α is an action of G on A, let $H^{\alpha}(A)$ denote the family of non-zero, globally α-invariant, hereditary C*-subalgebras of A, and let $H_B^{\alpha}(A)$ denote the subset of algebras $B \in H^{\alpha}(A)$ such that the closed ideal in A generated by B intersects every other non-zero ideal in A non-trivially. Define the <u>Connes spectrum</u> or <u>Γ-spectrum</u> of α as

(2.6.1) $\Gamma(\alpha) = \bigcap \{\mathrm{Spec}^{\alpha}(B) \mid B \in H^{\alpha}(A)\}$

and the <u>Borchers spectrum</u> of α as

(2.6.2) $\Gamma_B(\alpha) = \bigcap \{\mathrm{Spec}^{\alpha}(B) \mid B \in H_B^{\alpha}(A)\}$

Define A to be <u>G-simple</u> if A does not have any nontrivial closed, twosided, globally G-invariant ideals, and define A to be <u>G-prime</u> if any two such ideals intersect.

These concepts for von Neumann algebras goes back to Connes [Con 2], and they have been developed in the C*-framework in [Ped 1], and more recently in [OP 1]. Note that the condition $\Gamma(\alpha) = \hat{G}$ is a condition of freeness for the action G, while the conditions of G-simplicity and G-primeness are conditions of ergodicity. We have trivially $\Gamma(\alpha) \subseteq \Gamma_B(\alpha)$ and $\Gamma(\alpha) = \Gamma_B(\alpha)$ if A is G-prime. We need the following two basic facts

<u>Lemma 2.6.4</u>

$\Gamma(\alpha)$ is a closed subgroup of $\hat{G}$.

<u>Proof</u>

[Ped 1], Proposition 8.8.4.

<u>Lemma 2.6.5</u>

For any compact subset K of $\hat{G}$ and any neighborhood V of 0 in $\hat{G}$, there is a $B \in H_B^{\alpha}(A)$ such that

(2.6.3) $\mathrm{Spec}(\alpha|_B) \cap K \subseteq (\Gamma_B(\alpha) + V) \cap K$

Proof [OP 1], Lemma 3.6.

We are now ready to state and prove the announced classification result.

Theorem 2.6.6 (Kishimoto-Robinson)

Let A be a C*-algebra, G a locally compact abelian group and α an action of G on A such that

(2.6.4) A is α-prime,

and

(2.6.5) $\hat{G}/\Gamma(\alpha)$ is compact.

Let δ be a closed derivation on A such that

(2.6.6) $[\delta, \alpha] = 0$,

and

(2.6.7) $A_F^\alpha \subseteq D(\delta)$.

It follows that δ has a decomposition

$$\delta = \delta_0 + \tilde{\delta}$$

where δ_0 is the generator of a one-parameter subgroup of the action α of G, and $\tilde{\delta}$ is a bounded derivation. This decomposition is unique.

Proof

We follow the proof of Theorem 2.1 in [KR 1]. It follows from Theorem 2.6.2 that δ generates a strongly continuous one-parameter group β of *-automorphisms of A. As β commutes with α, we may define an action σ of $G \times \mathbb{R}$ on A by

$$\sigma_{(g,\, t)} = \alpha_g \beta_t$$

The Connes spectrum $\Gamma(\sigma)$ of σ is then a closed subgroup of $(G \times \mathbb{R})^{\wedge} = \hat{G} \times \mathbb{R}$ by Lemma 2.6.4.

Define π_1 as the canonical projection from $\hat{G} \times \mathbb{R}$ onto $\hat{G}$, and define $H = \pi_1(\Gamma(\sigma))$. If K is a compact subset of $\hat{G}$, then $\delta|_{A^\alpha(K)}$ is bounded and consequently $\mathrm{Spec}(\sigma) \cap (K \times \mathbb{R})$ $\subseteq K \times [-\| \delta|_{A^\alpha(K)} \|, +\| \delta|_{A^\alpha(K)} \|]$, i.e. $\mathrm{Spec}(\sigma) \cap (K \times \mathbb{R})$ is compact, [BR 1], Proposition 3.2.40. If in addition $K \cap H = \phi$, then $(\mathrm{Spec}(\sigma) \cap (K \times \mathbb{R}) \cap \Gamma(\sigma) = \phi$, and it follows from Lemma 2.6.5 that there is a $B \in H^\sigma(A) = H^\sigma_B(A)$ such that

$$\mathrm{Spec}\ (\sigma|_B) \cap (\mathrm{Spec}\ (\sigma) \cap (K \times \mathbb{R})) = \phi$$

i.e.

$$\mathrm{Spec}\ (\sigma|_B) \cap (K \times \mathbb{R}) = \phi$$

Thus $\mathrm{Spec}\ (\alpha|_B) \cap K = \phi$, which implies

(2.6.8) $\Gamma(\alpha) \subseteq H$,

(We note parenthetically that essential use of the ergodicity 2.6.4 were made to derive 2.6.8., consider the example $A = B \oplus B$, $G = \mathbb{R}$, $\alpha_t = \tau_t \oplus \tau_t$ where $\Gamma(\tau) = \mathbb{R}$ and $\beta_t = \tau_t \oplus 1$. Then $\sigma_{(s,\ t)} = \tau_{s+t} \oplus \tau_s$ and one verifies $\mathrm{Spec}\ (\sigma) = \{(x,\ x) |\ x \in \mathbb{R}\}$ $\cup \{(x,0) | x \in \mathbb{R}\}$ and $\Gamma(\sigma) = \{0\}$, hence $H = \{0\}$ and $H \not\supseteq \Gamma(\alpha) = \mathbb{R}$. This example also shows that 2.6.4 is essential for the conclusion of the theorem.)

Next note that as $\mathrm{Spec}\ (\sigma) \cap (K \times \mathbb{R})$ is compact for any compact $K \subset \hat{G}$, we have that $\Gamma(\sigma) \cap (\{0\} \times \mathbb{R})$ is a compact subgroup of $\{0\} \times \mathbb{R}$, thus $\Gamma(\sigma) \cap (\{0\} \times \mathbb{R}) = \{0\} \times \{0\}$. It follows that:

(2.6.9) π_1 is injective on $\Gamma(\sigma)$.

Thus, for each $\gamma \in H$, there is a unique $\rho(\gamma) \in \mathbb{R}$ with $(\gamma,\ \rho(\gamma)) \in \Gamma(\sigma)$. Since $\Gamma(\sigma)$ is a group, the map $\rho \colon \gamma \to \rho(\gamma)$ is an additive morphism from H into $\mathbb{R}$.

Define

$$(2.6.10)\quad I = \Gamma(\sigma)^{\perp} = \{(g,\ t) \in G \times \mathbb{R} \mid \gamma(g)e^{i\rho(\gamma)t} = 1 \text{ for all } \gamma \in H\}$$

If π_2 is the canonical projection from $G \times \mathbb{R}$ onto $\mathbb{R}$, then $\pi_2(I) = \mathbb{R}$ and $\ker \pi_2\big|_I = H^{\perp}$. But $H^{\perp}$ is discrete, since it is the dual of the group $\hat{G}/H$ which is compact by 2.6.5 and 2.6.8. Thus, if I_0 is the connected component of 0 in I, I_0 is isomorphic to $\mathbb{R}$. Define $g(t) \in G$ for $t \in \mathbb{R}$ by $(g(t),\ -t) \in I_0$. Then $t \to g(t)$ is a continuous homomorphism from $\mathbb{R}$ into G such that

$$(2.6.11)\quad \gamma(g(t)) = e^{i\rho(\gamma)t}$$

for $\gamma \in H$, by 2.6.10.

We next claim that $\mathrm{Spec}\ (\sigma)/I_0^{\perp}$ is bounded in $\hat{I}_0 \cong \mathbb{R}$. If it is not bounded there is a sequence $(\gamma_n,\ \rho_n) \in \mathrm{Spec}\ (\sigma)$ such that the characters

$$t \to \gamma_n(g(t))e^{-it\rho_n}$$

of $\mathbb{R}$ are unbounded. Since replacing $(\gamma_n,\ \rho_n)$ by $(\gamma_n,\ \rho_n) + (\gamma,\ \rho(\gamma))$ with $\gamma \in H$ does not change these characters, see 2.6.11, and $\hat{G}/H$ is compact, we can assume that γ_n varies over some compact subset K of $\hat{G}$, with $K + H = \hat{G}$. This shows that σ_n must be unbounded. Since $\mathrm{Spec}\ (\sigma) \cap (K \times \mathbb{R})$ is compact this is a contradiction.

Finally, define a one-parameter group τ of *-automorphisms of A by

$$(2.6.12)\quad \tau_t = \alpha_{g(t)}\, \beta_{-t}\ .$$

Then $\mathrm{Spec}(\tau)$ identifies with the closure of $\mathrm{Spec}(\sigma)/I_0^{\perp}$, and thus $\mathrm{Spec}(\tau)$ is a compact subset of $\mathbb{R}$. But this means that τ is uniformly continuous, and the generator $-\tilde{\delta}$ of τ is bounded, see [BR 1], Propositions 3.2.41 and 3.1.1., or the proof of Proposition 2.2.14 in the lecture notes. But differentiating 2.6.12 on A_F^{α}, we obtain

$$-\tilde{\delta}(x) = \delta_0(x) - \delta(x)$$

for $x \in A_F^\alpha$, where δ_0 is the generator of the one-parameter group $t \to \alpha_{g(t)}$.

The uniqueness of the decomposition follows by noting that I_0 is the only one-parameter subgroup of $\hat{G} \times \mathbb{R}$ which makes the closure of Spec $(\sigma)/I_0^{\perp}$ compact.

Example 2.6.7 (Kishimoto-Robinson)

We remarked already in the proof that simple examples show that Theorem 2.6.6 breaks down if A is not α-prime. But also the condition that $\hat{G}/\Gamma(\alpha)$ is compact is essential by the following example, which is worked out in detail in [KR 1]. Let $A = \bigotimes_{n=1}^{\infty} M_2$ be the infinite tensor product of the 2×2 matrix algebra M_2, see §1.3, and define representations $\beta, \alpha : \mathbb{R} \to \mathrm{Aut}\,(A)$ by

$$\beta_t = \otimes\, \beta_{n,t} \quad , \qquad \beta_{n,t} = \mathrm{Ad}\begin{pmatrix} 1 & 0 \\ 0 & e^{i\mu_n t} \end{pmatrix}$$

where μ_n is an arbtrary sequence of real numbers and

$$\alpha_t = \otimes\, \alpha_{n,t} \quad , \qquad \alpha_{n,t} = \mathrm{Ad}\begin{pmatrix} 1 & 0 \\ 0 & e^{i\lambda_n t} \end{pmatrix}$$

where λ_n is a sequence converging sufficiently fast to $+\infty$ (precisely $\lambda_n > 2^{n+1} M \sum_{k=1}^{n} |\mu_k|^{k=1}$ for $n = 1, 2, \ldots$ and $\lambda_n > 2M \sum_{k=1}^{n-1} \lambda_k$ for $n = 2, 3, \ldots$, where M is a positive constant, independent of β) If δ_β, δ_α are the generators of β, α, respectively, one can then show $D(\delta_\alpha) \subset D(\delta_\beta)$, but $D(\delta_\alpha) \neq D(\delta_\beta)$, and hence δ_β cannot be decomposed as $\lambda\delta_\alpha + \tilde{\delta}$ with $\lambda \in \mathbb{R}$ and $\tilde{\delta}$ bounded. In this case $\Gamma(\alpha) = \{0\}$, as expected from Theorem 2.6.6. (In fact $\Gamma(\beta)$ is the intersection of the closures of the sets $\{ \sum_{k=n}^{\infty} \varepsilon_k \mu_k |\ \varepsilon_k = \pm 1, 0,$ $\varepsilon_k = 0$ except for a finite number of $k\}$, $n = 1, 2, \ldots$, for a general β of the form above.

We will later give an example of an action of $\mathbb{T}$ on a simple, unital C^*-algebra A such that a general $\delta \in \mathrm{Der}(A_F^\alpha, A)$ does not even have a decomposition $\delta = \delta_0 + \tilde{\delta}$, where $\tilde{\delta}$ is approximately inner, see Example 2.7.10.

§2.6.3 Lie groups

Theorem 2.6.8 (Davies)

Let A be a C*-algebra, G a Lie group which is the semidirect product of an abelian Lie group by a compact Lie group, α an action of G on A, and δ a derivation such that

$$D(\delta) = A_\infty \quad \text{(See 2.2.3)}$$

and

$$[\delta, \alpha] = 0$$

It follows that δ is a pre-generator.

Proof

This is Theorem 3.2 in [Dav 1]. We give an outline of the proof. The hypothesis on G implies that G is unimodular and there exists a subset M of $L^1(G) \cap C^\infty(G)$ satisfying

(2.6.13) For all $f \in M$ there is a sequence $f_n \in C_{00}^\infty(G)$ such that

$$\lim_{n\to\infty} \|X_1 \dots X_m(f_n - f)\| = 0$$

for all finite sequences $X_1, \dots, X_m$ of right-invariant vector fields on G.

(2.6.14) M is an abelian subalgebra of $L^1(G)$

(2.6.15) $L^1(G)$ has an approximate identity consisting of positive functions in M

(2.6.16) For all $f \in M$ there exists a $h \in M$ with $h*f = f$

(2.6.17) M is closed under complex conjugation

(2.6.18) For all $f, g \in M$ there exists a $h \in M$ with

$$\int_G ds\, h(s)f(s^{-1}t)g(s^{-1}u) = f(t)g(u) \qquad \text{for all } t, u \in G$$

Actually, it is the existence of this subset M which is essential for the proof. Note first that δ is continuous in the Fréchet topology on A_∞ by Theorem 2.3.1. As a consequence we deduce that

$$\delta\alpha_f x = \alpha_f \delta x \tag{2.6.19}$$

for $x \in A_\infty$, first for $f \in C_{00}^\infty(G)$ and then for $f \in M$, using 2.6.13. (We cannot use 2.5.24 at this stage, as we do not know δ is closed.) For $f \in M$, define

$$A_f = \{x \in A_\infty \mid \alpha_f(x) = x\} \tag{2.6.20}$$

The subspaces A_f replaces the spectral subspaces $A^\alpha(\Lambda)$ in the proofs of Theorems 2.6.1 and 2.6.2. These subspaces A_f are norm closed, and one uses 2.6.14, 2.6.15 and 2.6.16 to show that these spaces form an incresing net with dense union in A. It next follows from 2.6.17 and 2.6.18 that $\bigcup\{A_f \mid f \in M\}$ is a *-algebra.

Now, it follows from 2.6.20 and 1.5.5 that the norm topology from A and the Fréchet topology from A_∞ coincides on the closed subspace A_f of A. As δ is continuous in the Fréchet topology, it follows that $\delta|_{A_f}$ is bounded. But $\delta(A_f) \subseteq A_f$ by 2.6.19, and hence δ exponentiates by power series expansion on A_f. As δ is a *-derivation, $e^{t\delta}$ is a *-automorphism of $\bigcup\{A_f \mid f \in M\}$. One now uses Lemma 2.5.3 as in the proofs of Theorems 2.6.1 and 2.6.2 to verify that each $e^{t\delta}$ is an isometry, and hence δ extends by closure to a generator.

<u>Remark 2.6.9</u>

Note that the hypotheses of Theorem 2.6.8 imply that $\delta \in \mathrm{Der}(A_\infty, A_\infty)$, as a consequence of 2.6.19 and Theorem 2.2.3.

<u>Problem 2.6.10</u> (Davies)

Can one find a subset M of $L^1(G)$ with the properties 2.6.13 -18 for a general Lie group G?

§2.7 Cocycles and derivations

Let α be an action of a compact or abelian locally compact group G on a C*-algebra A. In analyzing derivations $\delta \in \mathrm{Der}\,(A_F^\alpha, A)$, (see Definitions 2.2.13 and 2.2.15), two main techniques are used

1. Fourier analysis of the map

$$g \in G \to \alpha_g \delta \alpha_g^{-1} .$$

2. Decompositions of δ of the form

$$\delta = \delta_0 + \tilde{\delta}$$

where $\tilde{\delta}$ is bounded and $\delta_0\big|_{A^\alpha} = 0$. The last technique is of course mostly relevant when G is compact. Hence it is of interest of analyzing derivations δ with the property $\delta\big|_{A^\alpha} = 0$, where A^α is the fixed point algebra of a compact action. These derivations can be characterized by certain cocycles over $\hat{G}$, as we shall see.

§2.7.1 Abelian compact groups

Slightly abusing terminology, we define:

Definition 2.7.1

Let α be an action of a topological group G on a C*-algebra A, and let π be a (non-degenerate) representation of A on a Hilbert space. The representation π is said to be G-covariant if α extends to a σ-weakly continuous action $\hat{\alpha}$ of G on the σ-weak closure $\pi(A)''$ of A, i.e. if there exists a point-σ-weakly continuous map $g \in G \to \hat{\alpha}_g \in \mathrm{Aut}\,(\pi(A)'')$ such that

$$(2.7.1) \quad \hat{\alpha}_g(\pi(x)) = \pi(\alpha_g(x))$$

for all $g \in G$, $x \in A$.

Up to multiplicity, this means that $\hat{\alpha}$ is implemented by a strongly continuous unitary representation of G, [BR 1], Corollary 2.5.32.

Let G be a compact group and α an action of G on a C*-algebra A covariantly and faithfully represented on a Hilbert space H, and put $M = A''$. The action α extends to an action of G on M, and the spectral spaces $A_1^\alpha(\gamma)$ are σ-weakly dense in $M_1^\alpha(\gamma)$ since the maps $P_{ij}(\gamma) = \int_G dg\ d(\gamma)\overline{\gamma_{ji}(g)}\alpha_g$ are σ-weakly continuous, [BR 1], Proposition 3.1.4. Since $M^\alpha(\gamma)$ is an M^α-module, the projections $E(\gamma) = [M^\alpha(\gamma) H]$ onto the closed linear span of $M^\alpha(\gamma) H$ is contained in the center of M^α(Alternatively, $E(\gamma)$ is the range-projection of the ideal $M_1^\alpha(\gamma)M_1^\alpha(\gamma)^*$ in M^α). Let

(2.7.2) $C(\gamma)$ = the relative commutant of $M^\alpha E(\gamma)$ in $E(\gamma)ME(\gamma)$,

(2.7.3) $\mathcal{D}(\gamma)$ = the relative commutant of $A_1^\alpha(\gamma)\, A_1^\alpha(\gamma)^*$ in the multiplier algebra of $\overline{A^\alpha(\gamma)\ A\ A^\alpha(\gamma)^*}^{\|\ \|}$.

Then clearly $\mathcal{D}(\gamma) \subseteq C(\gamma)$ for all $\gamma \in \hat{G}$.

Lemma 2.7.2

Assume that the compact group G is abelian. For each $a \in C(-\gamma)$ there exists a unique element $\beta_\gamma(a) \in C(\gamma)$ such that

(2.7.4) $\beta_\gamma(a)x = xa$

for all $x \in A^\alpha(\gamma)$. The map β_γ is a *-isomorphism from $C(-\gamma)$ onto $C(\gamma)$, and β_γ maps $\mathcal{D}(-\gamma)$ onto $\mathcal{D}(\gamma)$. If $E(\gamma) = \mathbb{1}$ for all $\gamma \in \hat{G}$, then $\gamma \in \hat{G} \to \beta_\gamma$ is a representation of $\hat{G}$ in Aut $(C(0))$.

Remark 2.7.3

If $M^\alpha(\gamma)$ contains a unitary operator u, one has $E(\gamma) = E(-\gamma) = \mathbb{1}$ and $\beta_\gamma(a) = u\,a\,u^*$ for $a \in C(\gamma) = C(0)$.

Proof

This lemma is a minor generalization of Lemma 1.5 in [BJKR 1], see [Bra 3], Lemma 4.4, [BGJ 1], Lemma 2.1 and [BG 1], remark before

Lemma 3.1.

Assume that $a \in C(-\gamma)$ is given. As $E(\gamma)$ is the range projection of the subspace $[A^{\alpha}(\gamma) H]$, the operator b defined on this subspace by

$$(2.7.5) \quad b(\sum_i x_i \xi_i) = \sum_i x_i a \xi_i$$

for $x_i \in A^{\alpha}(\gamma)$, $\xi_i \in H$ is unique whenever the definition is consistent. In fact, we will show that

$$(2.7.6) \qquad \|\sum_i x_i a \xi_i\| \leq \|a\| \; \|\sum_i x_i \xi_i\| ,$$

and hence the definition is indeed consistent, and $\|b\| \leq \|a\|$.

To prove 2.7.6, assume first that the finite sum contains only one element. Then

$$\begin{aligned} \|xa\xi\|^2 &= (a\xi, x^*xa\xi) \\ &= (a^*a\xi, x^*x\xi) \\ &= (a^*a\xi, |x|^2\xi) \\ &= (a^*a|x|\xi, |x|\xi) \\ &= \|a|x|\xi\|^2 \leq \|a\|^2 \; \||x|\xi\|^2 = \|a\|^2 \; \|x\xi\| \end{aligned}$$

where we have used x^*x, $|x| \in A^{\alpha}E(-\gamma)$, and $a \in C(-\gamma)$, see 2.2.30 and 2.7.2. To derive the inequality 2.7.6 for general finite sums one employs the previous reasoning on the matrices

$$X = \begin{bmatrix} x_1 & \dots & x_n \\ 0 & \dots & 0 \\ \cdot & & \cdot \\ \cdot & & \cdot \\ \cdot & & \cdot \\ 0 & & 0 \end{bmatrix}, \qquad A = \begin{bmatrix} a & 0 & \dots & 0 \\ 0 & a & \dots & 0 \\ \cdot & \cdot & & \cdot \\ \cdot & \cdot & & \cdot \\ \cdot & \cdot & & \cdot \\ 0 & 0 & \dots & a \end{bmatrix}, \qquad \tilde{\xi} = \begin{bmatrix} \xi_1 \\ \cdot \\ \cdot \\ \cdot \\ \xi_n \end{bmatrix}.$$

Next we show that $b \in E(\gamma)ME(\gamma)$. By Lemma 2.5.1, $A^{\alpha}(\gamma)\, A^{\alpha}(\gamma)^*$ has an approximate identity e_{δ} of the form

$$e_\delta = \sum_i x_i^\delta x_i^{\delta *}$$

where $x_i^\delta \in A^\alpha(\gamma)$. From the defining relation 2.7.5 it follows that

$$(2.7.7) \quad be_\delta = \sum_i x_i^\delta a x_i^{\delta *}$$

Since $\lim_\delta e_\delta = E(\gamma)$, where the limit exists in the strong operator topology, it follows that

$$(2.7.8) \quad b = \lim_\delta \sum_i x_i^\delta a x_i^{\delta *} \in E(\gamma) M E(\gamma) .$$

Now we show that b commutes with $A^\alpha(\gamma)\, A^\alpha(\gamma)^*$, and thus with its weak closure $M^\alpha E(\gamma)$. If $c \in A^\alpha(\gamma)\, A^\alpha(\gamma)^*$, we have for all $x_i \in A^\alpha(\gamma)$, $\xi_i \in H$ that $cx_i \in A^\alpha(\gamma)$, and hence

$$\begin{aligned} cb(\sum_i x_i \xi_i) &= c \sum_i x_i a \xi_i \\ &= \sum_i (cx_i) a \xi_i \\ &= b \sum_i cx_i \xi_i = bc \sum_i x_i \xi_i \end{aligned}$$

It follows that b commutes with $A^\alpha(\gamma)\, A^\alpha(\gamma)^*$, and hence $b \in C(\gamma)$.

Note that if $a \in \mathcal{D}(-\gamma)$, it follows from 2.7.7 that

$$be_\delta c \in \overline{A^\alpha(\gamma)\ A\ A^\alpha(\gamma)^*}^{\|\ \|}$$

for all $c \in A^\alpha(\gamma)\ A\ A^\alpha(\gamma)^*$, and as e_δ is an approximate identity for the latter algebra by Lemma 2.5.2, it follows that b is a left multiplier of $\overline{A^\alpha(\gamma)\ A\ A^\alpha(\gamma)^*}^{\|\ \|}$. Replacing a by a^*, b is replaced by b^* as we shall see in a moment (proving that β_γ is a $*$-isomorphism) and hence b^* a left multiplier, i.e. b is a two-sided multiplier. This shows that $b \in \mathcal{D}(\gamma)$ if $a \in \mathcal{D}(-\gamma)$.

Now define $\beta_\gamma\colon C(-\gamma) \to C(\gamma)$ by $\beta_\gamma(a) = b$

That β_γ is a morphism follows from the defining relation 2.7.5:

$$\alpha_\gamma(a_1a_2)x = xa_1a_2 = \alpha_\gamma(a_1)xa_2 = \alpha_\gamma(a_1)\alpha_\gamma(a_2)x \ ,$$

for all $a_1, a_2 \in C(-\gamma)$, $x \in A^\alpha(\gamma)$.

We next show that α_γ is a *-morphism, i.e. $\alpha_\gamma(a^*) = \alpha_\gamma(a)^*$ for $a \in C(-\gamma)$. Put $b = \alpha_\gamma(a)$, $b_1 = \alpha_\gamma(a^*)$. We must show that $b_1 = b^*$. But if $x, y \in A^\alpha(\gamma)$, $\xi, \eta \in H$ we have

$$(b_1x\xi, y\eta) = (xa^*\xi, y\eta) = (a^*\xi, x^*y\eta) = (\xi, x^*ya\eta) = (x\xi, ya\eta) = (x\xi, by\eta)$$

where we used $x^*y \in A^\alpha(-\gamma)\ A^\alpha(-\gamma)^*$ and hence $ax^*y = x^*ya$. It follows that $b_1 = b^*$.

We thus know that $\beta_\gamma : C(-\gamma) \to C(\gamma)$ is a *-morphism. But since $A^\alpha(-\gamma) = A^\alpha(\gamma)^*$, it follows from the adjoint relation $x^*b^* = a^*x^*$ that $\beta_{-\gamma}$ is the inverse of β_γ. Therefore β_γ is a *-isomorphism.

If $E(\gamma) = 1$ for all $\gamma \in \hat{G}$, it follows from $A^\alpha(\gamma_1)\ A^\alpha(\gamma_2) \subseteq A^\alpha(\gamma_1 + \gamma_2)$ and 2.7.8 that

$$\beta_{\gamma_1}\beta_{\gamma_2} = \beta_{\gamma_1+\gamma_2}$$

i.e. β is a representation of $\hat{G}$ in Aut $(C(0))$.

Now, if δ is a derivation of A such that $A^\alpha \subseteq D(\delta)$ and $\delta|_{A^\alpha} = 0$, then $D(\delta)$ is an A^α-module under left and right multiplication, and δ is a module map i.e.

$$\delta(ax) = a\delta(x), \quad \delta(xa) = \delta(x)a$$

for all $a \in A^\alpha$, $x \in D(\delta)$. This is still true if δ is a dissipation rather than a derivation.

Lemma 2.7.4 (Evans)

Let δ be a dissipation of a C*-algebra A (see Definition 1.1.1) and let $\mathcal{D}$ be a *-subalgebra of $D(\delta)$ such that

(2.7.9) $\delta|_{\mathcal{D}} = 0$

It follows that

(2.7.10) $\delta(ax) = a\delta(x), \quad \delta(xa) = \delta(x)a$

for all $a \in \mathcal{D}$, $x \in D(\delta)$.

Proof

This is Lemma 5.3 in [BEv 1]. If ω is a state on A, the dissipation property of δ implies that

$$D(x, y) = \omega(\delta(x)^*y + x^*\delta(y) - \delta(x^*y))$$

is a non-negative sesquilinear form on $D(\delta) \times D(\delta)$. If $a \in \mathcal{D}$, then a^*, $a^*a \in \mathcal{D}$ and 2.7.9 implies

$$D(a, a) = 0 .$$

But then Schwarz's inequality for D implies that

$$D(a^*, x) = 0 = D(x^*, a)$$

whenever $a \in \mathcal{D}$, $x \in D(\delta)$. Since this is true for all states ω we get

$$a\delta(x) - \delta(ax) = 0 = \delta(x)a - \delta(xa) .$$

The preceding lemma motivates the following.

Lemma 2.7.5

Let α be an action of an abelian compact group G on a C^*-algebra A covariantly and faithfully represented on a Hilbert space H. Let δ: $A_F^\alpha \to M = A''$ be a *-linear map which is a left and right A^α-module map:

(2.7.11) $\delta(ax) = a\delta(x)$, $\delta(xa) = \delta(x)a$

for all $x \in A_F^\alpha$, $a \in A^\alpha$. It follows that for each $\gamma \in \hat{G}$ there exists an operator $L(\gamma)$ in the von Neumann algebra $C(\gamma)$ (see 2.7.2), and an operator $R(\gamma)$ in $C(-\gamma)$ such that

(2.7.12) $\delta(x) = L(\gamma)x = xR(\gamma)$

for all $x \in A^{\alpha}(\gamma)$. We have the relations

(2.7.13) $L(\gamma) = R(-\gamma)^*$

(2.7.14) $L(\gamma) = \beta_{\gamma}(R(\gamma))$,

where $\beta_{\gamma}: C(-\gamma) \to C(\gamma)$ is the *-isomorphism defined in Lemma 2.7.2.

If δ is a derivation then

(2.7.15) $L(\gamma) = -L(\gamma)^*$

for all γ, and if $E(\gamma_1, \gamma_2)$ denotes the projection

(2.7.16) $E(\gamma_1, \gamma_2) = E(\gamma_1)\beta_{\gamma_1}(E(-\gamma_1)E(\gamma_2))$

then L satisfies the partial cocycle relation

(2.7.17) $L(\gamma_1+\gamma_2)E(\gamma_1,\gamma_2) = L(\gamma_1)E(\gamma_1,\gamma_2) + \beta_{\gamma_1}(E(-\gamma_1)L(\gamma_2))E(\gamma_1,\gamma_2)$

and the skew-adjointness relation:

(2.7.18) $L(-\gamma) = \beta_{-\gamma}(L(\gamma)^*)$

If the A^{α}-module map δ maps A_F^{α} into A, then $L(\gamma)$ is a left multiplier of $\overline{A^{\alpha}(\gamma)\, A\, A^{\alpha}(\gamma)^*}^{\|\ \|}$ and $R(-\gamma)$ is a right multiplier of this algebra. Thus, if δ in addition is a derivation:

(2.7.19) $L(\gamma) \in \mathcal{D}(\gamma)$, $R(\gamma) \in \mathcal{D}(-\gamma)$

see 2.7.3.

If conversely $\gamma \to L(\gamma) \in C(\gamma)$ is a map satisfying 2.7.17 and 2.7.18, then the map $\delta: A_F^{\alpha} \to M$ defined by 2.7.12 is a *-derivation. If in addition 2.7.19 holds, the so defined δ maps A_F^{α} into A.

Remark 2.7.6

This Lemma was developed succesively in [BJ 1], proof of Theorem 5.1, [BEv 1], Lemma 5.4, [BJKR 1], Lemma 1.2, [LP 1], Lemma 3.5,

[BGJ 1], Proposition 2.3 and [BG 1], Lemma 3.1. Note that the technical assumption $E(\gamma)\delta(x) = \delta(x)$ for all $x \in A^{\alpha}(\gamma)$ which is made in the last reference is unnecessary, it follows from the other assumptions.

We first prove the following general lemma.

Lemma 2.7.7

Let α be an action of a compact group G on a C*-algebra A covariantly and faithfully represented on a Hilbert space H, with weak closure M. If $x = [x_{ij}]_{i=1}^{n}{}_{j=1}^{d(\gamma)}$ is an element in $M_n^{\alpha}(\gamma)$, and $x = (xx^*)^{\frac{1}{2}}u$ is its polar decomposition as an element in $L(H^{d(\gamma)}, H^n)$, then

$$(2.7.20)\quad (xx^*)^{\frac{1}{2}} \in M^{\alpha} \otimes M_n \quad , \quad u \in M_n^{\alpha}(\gamma) .$$

Furthermore, if $x \in A_n^{\alpha}(\gamma)$, then

$$(2.7.21)\quad (xx^*)^{\frac{1}{4}} \in A^{\alpha} \otimes M_n \quad , \quad (xx^*)^{\frac{1}{4}}u \in A_n^{\alpha}(\gamma) ,$$

i.e. x has the form $x = ay$ where $a \in \overline{A_n^{\alpha}(\gamma)\, A_n^{\alpha}(\gamma)^*}^{\|\ \|}$, $y \in A_n^{\alpha}(\gamma)$.

Proof

We follow [BG 1], Observation 3. First note that 2.2.42 implies that $\alpha_g(xx^*) = xx^*$, i.e. $xx^* \in M^{\alpha} \otimes M_n$ and hence $(xx^*)^{\frac{1}{2}}$, $(xx^*)^{\frac{1}{4}} \in M^{\alpha} \otimes M_n$ (and these elements are contained in $\overline{A_n^{\alpha}(\gamma)\, A_n^{\alpha}(\gamma)^*}^{\|\ \|}$ if $x \in A$).

Now, on one hand

$$\alpha_g(x) = x(\mathbb{1} \otimes \gamma(g)) = (xx^*)^{\frac{1}{2}}\, u\, (\mathbb{1} \otimes \gamma(g))$$

and on the other hand

$$\alpha_g(x) = \alpha_g(xx^*)^{\frac{1}{2}}\alpha_g(u) = (xx^*)^{\frac{1}{2}}\alpha_g(u)$$

Comparing these two expressions and using the uniqueness of the polar decomposition, we obtain

$$\alpha_g(u) = u(1 \otimes \gamma(g)),$$

i.e. $u \in M_n^\alpha(\gamma)$.

If $x \in A^\alpha(\gamma)$, then

$$(xx^*)^{\frac{1}{4}}u = (xx^*)^{-\frac{1}{4}}x = \lim_{m\to\infty} (a_m x)$$

where

$$a_m = f_m(xx^*)$$

and

$$f_m(t) = \begin{cases} t^{-\frac{1}{4}} & \text{for } t \geq \frac{1}{m} \\ tm^{\frac{5}{4}} & \text{for } 0 \leq t \leq \frac{1}{m} . \end{cases}$$

Then $a_m \in A^\alpha \otimes M_n$ for all m, and the limit exists in norm by the estimate

$$||a_m x - (xx^*)^{\frac{1}{4}} u|| = || f_m(xx^*)(xx^*)^{\frac{1}{2}} - (xx^*)^{\frac{1}{4}}||$$

$$\leq \sup\{|f_m(t)t^{\frac{1}{2}} - t^{\frac{1}{4}}| \,\big|\, t > 0\} < (\tfrac{1}{m})^{\frac{1}{4}} \xrightarrow[m\to\infty]{} 0 .$$

Thus $(xx^*)^{\frac{1}{4}}u \in A$.

<u>Proof of Lemma 2.7.5</u>

We mainly follow the proof of Lemma 3.1 in [BG 1]. By Lemma 2.5.1, $A^\alpha(\gamma)\,A^\alpha(\gamma)^*$ has an approximate identity of the form

$$e_\tau = \sum_i a_i^\tau a_i^{\tau *}$$

where $a_i^\tau \in A^\alpha(\gamma)$ and each sum is finite. Then $\lim_{\tau\to\infty} e_\tau = E(\gamma)$ strongly on H. Define

$$(2.7.22) \quad L_\tau(\gamma) = \sum_i \delta(a_i^\tau)a_i^{\tau *}$$

If $x \in A^{\alpha}(\gamma)$, then

$$
\begin{aligned}
L_{\tau}(\gamma)x &= \sum_i \delta(a_i^{\tau})a_i^{\tau *}x \\
&= \sum_i \delta(a_i^{\tau}\, a_i^{\tau *}\, x) \\
&= \sum_i a_i^{\tau}\, a_i^{\tau *}\, \delta(x) \\
&= e_{\tau}\delta(x)
\end{aligned}
\tag{2.7.23}
$$

where we used 2.7.11 and the facts that $a_i^{\tau *}x$, $a_i^{\tau}\, a_i^{\tau *} \in A^{\alpha}$. But Lemma 2.7.7 implies that x has the form $x = ay$ where $a \in \overline{A^{\alpha}(\gamma)\; A^{\alpha}(\gamma)^*}^{\|\ \|}$ and $y \in A^{\alpha}(\gamma)$. Thus $\delta(x) = a\delta(y)$, and the following limit exists in norm

$$
\lim_{\tau\to\infty} L_{\tau}(\gamma)x = \lim_{\tau\to\infty} e_{\tau}a\delta(y) = a\delta(y) = \delta(x)
\tag{2.7.24}
$$

Hence we may define an operator $L_0(\gamma)$ from $E(\gamma)H$ into $E(\gamma)H$ with domain

$$
D(L_0(\gamma)) = A^{\alpha}(\gamma)H
$$

by

$$
L_0(\gamma)(\sum_i x_i\xi_i) = \lim_{\tau\to\infty} L_{\tau}(\gamma)(\sum_i x_i\xi_i) = \sum_i \delta(x_i)\xi_i
\tag{2.7.25}
$$

for $x_i \in A^{\alpha}(\gamma)$, $\xi_i \in H$.

We next show that $L_0(\gamma)$ is closable by showing that $L_0(\gamma)^*$ is densely defined on $E(\gamma)H$. Actually $A^{\alpha}(\gamma)\; A^{\alpha}(\gamma)^*H \subseteq D(L_0(\gamma)^*)$ by the computation

$$
\begin{aligned}
(L(\gamma)x\xi,\ yz^*\eta) &= (\delta(x)\xi,\ yz^*\eta) = (zy^*\delta(x)\xi,\ \eta) = (\delta(zy^*x)\xi,\ \eta) \\
&= (\delta(z)y^*x\xi, \eta) = (x\xi,\ y\delta(z^*)\eta)
\end{aligned}
$$

vaild for $x, y, z \in A^{\alpha}(\gamma)$, $\xi, \eta \in H$. Hence $yz^*\eta \in D(L_0(\gamma)^*)$ and

$$
L_0(\gamma)^*yz^*\eta = y\delta(z^*)\eta
\tag{2.7.26}
$$

As $e_\tau H \subseteq A^\alpha(\gamma)A^\alpha(\gamma)^*H$ and $e_\tau \to E(\gamma)$ strongly, it follows that $L_0(\gamma)^*$ is densely defined. Thus $L_0(\gamma)$ is closable, and we denote the closure by $L(\gamma)$.

If $x \in A^\alpha(\gamma)$, $a \in A^\alpha$ and $\xi \in H$ we have

$$L(\gamma)ax\xi = \delta(ax)\xi = a\delta(x)\xi = aL(\gamma)x\xi$$

and by closure the relation

$$L(\gamma)a\xi = aL(\gamma)\xi$$

holds for all $a \in M^\alpha = A^{\alpha''}$ and $\xi \in D(L(\gamma))$, i.e. $L(\gamma)$ is affiliated to $(M^\alpha)'E(\gamma)$.

Since $L(\gamma)$ is a limit of the operators $L_\tau(\gamma)$ on the core $A^\alpha(\gamma)H$, and $L_\tau(\gamma) = \sum_i \delta(a_i^\tau)a_i^{\tau*} \in E(\gamma)ME(\gamma)$ by Lemma 2.7.7, it follows that $L(\gamma)$ is also affiliated to $E(\gamma)ME(\gamma)$, thus $L(\gamma)$ is affiliated to $C(\gamma) = E(\gamma)ME(\gamma) \cap (M^\alpha)'E(\gamma)$.

We now argue that $\delta\big|_{A^\alpha(\gamma)}$ is bounded. If $x_n \in A^\alpha(\gamma)$ is a sequence such that $x_n \to 0$ and $\delta(x_n) \to y$, then, for all $\xi \in H$, $x_n\xi \to 0$ while $L(\gamma)x_n\xi = \delta(x_n)\xi \to y\xi$. As $L(\gamma)$ is closed, it follows that $y\xi = 0$ and so $y = 0$. Thus $\delta\big|_{A^\alpha(\gamma)}$ is closed and therefore bounded. (We could have used Theorem 2.3.8.)

Next we observe that $M^\alpha(\gamma)H \subset D(L(\gamma))$ and that $L(\gamma)M^\alpha(\gamma) \subseteq M$. Let $x \in M^\alpha(\gamma)$. Using the Kaplansky density theorem and applying the σ-weakly continuous projection $P(\gamma) = \int_G dg\,\overline{\langle\gamma,g\rangle}\,\alpha_g$ from M onto $M^\alpha(\gamma)$, we can find a net $x_\tau \in A^\alpha(\gamma)$ such that $\|x_\tau\| \leq \|x\|$ and x_τ converges weakly to x. Considering the boundedness of δ on $A^\alpha(\gamma)$ and the weak compactness of the unit ball of M, and passing to a subnet of x, we can suppose that $\delta(x_\tau)$ converges to some $y \in M$. But then as $L(\gamma)$ is closed with respect to the weak topology on H and $\delta(x_\tau) = L(\gamma)x_\tau$ for all τ, it follows by limiting that $xH \subseteq D(L(\gamma))$ and $y = L(\gamma)x$.

We can now define an operator δ from M_F^α into M by

(2.7.27) $\delta(x) = L(\gamma)x$

for all $x \in M^{\alpha}(\gamma)$. As $L(\gamma)$ is closed, it follows as above that δ is σ-weakly closed and hence σ-weakly continuous on each spectral subspace $M^{\alpha}(\gamma)$. The operator δ extends our original δ, and it follows by taking limits that the extended δ is a *-derivation.

Our next aim is to show that the operator $L(\gamma)$ affiliated to $C(\gamma)$ is bounded, and thus actually included in $C(\gamma)$.

If $x \in M^{\alpha}(\gamma)$, and $x = u|x|$ is its polar decomposition, then $u \in M^{\alpha}(\gamma)$ and $|x| \in M^{\alpha}$ by Lemma 2.7.7. If $\{u_{\tau}\}$ is a family of partial isometries in $M^{\alpha}(\gamma)$ with mutually orthogonal initial projections $u_{\tau}^{*}u_{\tau} \leq E(-\gamma)$ and mutually orthogonal range projections $u_{\tau}u_{\tau}^{*} \leq E(\gamma)$, then $\omega = \sum_{\tau} u_{\tau}$ is a partial isometry in $M^{\alpha}(\gamma)$, and by Zorn's lemma, there exists a maximal partial isometry in $M^{\alpha}(\gamma)$. Maximality is characterized by the property

$$(E(\gamma) - uu^{*})M^{\alpha}(\gamma)(E(-\gamma) - u^{*}u) = 0,$$

since if this space is not zero, then it contains a partial isometry which can be added to u, and on the other hand, if w is a partial isometry properly containing u, then $(E(\gamma) - uu^{*})w(E(-\gamma) - u^{*}u) \neq 0$.

Next we show that if u is a maximal partial isometry in $M^{\alpha}(\gamma)$, then there exists a projection E in the center of $M^{\alpha}E(\gamma)$ with the property

$$uu^{*}(E(\gamma) - E) = (E(\gamma) - E)$$

and

$$u^{*}u\beta_{-\gamma}(E) = \beta_{-\gamma}(E) .$$

This can be seen as follows: Let E be the central support of the projection $E(\gamma)-uu^{*}$ in $M^{\alpha}E(\gamma)$. Then clearly $uu^{*}(E(\gamma)-E) = (E(\gamma)-E)$, while $uu^{*}P \lneq P$ whenever $P \leq E$ is a non-zero projection in the center of M^{α}. Suppose now that $\beta_{-\gamma}(E)(E(-\gamma) - u^{*}u) \neq 0$; then

$$\begin{aligned} 0 &\neq M^{\alpha}(\gamma)\beta_{-\gamma}(E)(E(-\gamma) - u^{*}u) \\ &= EM^{\alpha}(\gamma)\beta_{-\gamma}(E)(E(-\gamma) - u^{*}u) . \end{aligned}$$

If x is a non-zero element in the latter space, then

$$x = uu^*x + (E(\gamma) - uu^*)x$$

$$= uu^*Ex + (E(\gamma) - uu^*)Ex,$$

and hence either

1. $(E(\gamma) - uu^*)Ex \neq 0$, or

2. $uu^*Ex \neq 0$.

In case 1, the condition $(E(\gamma) - uu^*)M^{\alpha}(\gamma)(E(-\gamma) - u^*u) = 0$ is contradicted. In case 2, as $uu^*P \lneq P$ for each central projection $P \leq E$ in M^{α}, there is an element $y \in M^{\alpha}$ with

$$0 \neq (E(\gamma) - uu^*)yuu^*Ex$$

$$= (E(\gamma) - uu^*)yuu^*Ex(E(-\gamma) - u^*u),$$

and since $yuu^*Ex \in M^{\alpha}(\gamma)$, this again contradicts the maximality of u. This contradiction shows that

$$u^*u\beta_{-\gamma}(E) = \beta_{-\gamma}(E).$$

It follows from this and the other identity

$$uu^*(E(\gamma) - E) = E(\gamma) - E$$

that any $x \in M^{\alpha}(\gamma)$ has the decomposition

$$x = (E(\gamma) - E)x + Ex$$

$$= (E(\gamma) - E)x + x\beta_{-\gamma}(E)$$

$$= (E(\gamma) - E)uu^*x + xu^*u\beta_{-\gamma}(E)$$

and as u^*x and xu^* lie in M^{α}, we get

$$L(\gamma)x = \delta(x)$$

$$(2.7.28) \quad = (E(\gamma) - E)\delta(u)u^*x + xu^*\delta(u)\beta_{-\gamma}(E)$$

$$= [(E(\gamma) - E)\delta(u)u^* + \beta_\gamma(u^*\delta(u))E]\cdot x$$

It follows that $L(\gamma) = (E(\gamma) - E)\delta(u)u^* + \beta_\gamma(u^*\delta(u))E$, and $L(\gamma)$ is a bounded element in $C(\gamma) = E(\gamma)ME(\gamma) \cap (M^{\alpha}_{\gamma})'E(\gamma)$.

One can now define $R(\gamma)$ by

$$R(\gamma) = \beta_{-\gamma}(L(\gamma))$$

and the relations 2.7.12 and 2.7.14 follows from Lemma 2.7.2. If $x \in M^{\alpha}(\gamma)$, then as δ is a *-linear map and $M^{\alpha}(\gamma)^* = M^{\alpha}(-\gamma)$: $L(\gamma)x = \delta(x) = \delta(x^*)^* = (x^*R(-\gamma))^* = R(-\gamma)^*x$ and 2.7.13 follows.

If δ is a derivation, then the derivation property

$$(2.7.29) \quad \delta(x_1x_2) = \delta(x_1)x_2 + x_1\delta(x_2)$$

applied to $x_1 \in M^{\alpha}(\gamma_1)$, $x_2 \in M^{\alpha}(\gamma_2)$, gives

$$L(\gamma_1 + \gamma_2)x_1x_2 = L(\gamma_1)x_1x_2 + x_1L(\gamma_2)x_2$$

$$= (L(\gamma_1) + \beta_{\gamma_1}(E(-\gamma_1)L(\gamma_2)))x_1x_2$$

and as $E(\gamma_1, \gamma_2)$ is the range projection of $[M^{\alpha}(\gamma_1)M^{\alpha}(\gamma_2)H]$, 2.7.17 follows. Conversely, 2.7.17 implies 2.7.29, i.e. 2.7.17 implies that δ is a derivation. It is also easy to see that 2.7.18 is equivalent to the property.

$$\delta(x^*) = \delta(x)^*$$

for $x \in M^{\alpha}(\gamma)$. The relation 2.7.15, $L(\gamma) = -L(\gamma)^*$ follows by combining 2.7.18 with 2.7.17 applied to $\gamma_1 = \gamma$, $\gamma_2 = -\gamma$.

Now, if δ is an A^{α}-module map from A^{α}_{F} into A, it follows from 2.7.22 and Lemma 2.7.7 that $L_\tau(\gamma) \in \overline{A^{\alpha}(\gamma)\, A\, A^{\alpha}(\gamma)}^{\|\ \|}$, and as the limit 2.7.24 exists in norm for $x \in A^{\alpha}(\gamma)$, we deduce that

$$L(\gamma)\, A^{\alpha}(\gamma) \subseteq \overline{A^{\alpha}(\gamma) A}^{\|\ \|}$$

and hence

$$L(\gamma)\, \overline{A^{\alpha}(\gamma)\ A\ A^{\alpha}(\gamma)^*}^{\|\ \|} \subseteq \overline{A^{\alpha}(\gamma)\ A\ A^{\alpha}(\gamma)^*}^{\|\ \|}$$

i.e. $L(\gamma)$ is a left multiplier of $\overline{A^{\alpha}(\gamma)\ A\ A^{\alpha}(\gamma)^*}^{\|\ \|}$. Thus $L(\gamma)^* = R(-\gamma)$ is a right multiplier of this algebra. If δ in addition is a derivation it follows from the skewadjointness 2.7.15 of $L(\gamma)$ that $L(\gamma)$ is a two-sided multiplier, i.e. $L(\gamma) \in \mathcal{D}(\gamma)$. Thus $R(\gamma) = \beta_{-\gamma}(L(\gamma)) \in \mathcal{D}(-\gamma)$.

This ends the proof of Lemma 2.7.5.

<u>Example 2.7.8</u> ([BGJ 1], Proposition 2.3 and Theorem 3.1)

Suppose that the action α of the compact abelian group G has the property that

(2.7.30) $$\overline{A^{\alpha}(\gamma)\ A^{\alpha}(\gamma)^*}^{\|\ \|} = A^{\alpha}$$

for all $\gamma \in \hat{G}$. Then

(2.7.31) $\mathcal{D}(\gamma) = \mathcal{D}$ = the relative commutant of A^{α} in $M(A)$

for each $\gamma \in \hat{G}$. It follows from Lemma 2.7.5 that it is a one-one correspondence between derivations δ satisfying

(2.7.32) $$D(\delta) = A_F^{\alpha}\ , \qquad \delta\big|_{A^{\alpha}} = 0$$

and maps $\gamma \to L(\gamma) \in \mathcal{D} = M(A) \cap (A^{\alpha})'$ satisfying

(2.7.33) $$L(\gamma_1 + \gamma_2) = L(\gamma_1) + \beta_{\gamma_1}(L(\gamma_2))$$

(2.7.34) $$L(-\gamma) = \beta_{-\gamma}(L(\gamma)^*)$$

where β is defined in Lemma 2.7.2. This correspondence is given by

(2.7.35) $$\delta(x) = L(\gamma)x,$$

for $x \in A^{\alpha}(\gamma)$. The relation 2.7.33 expresses that L is an (additive) <u>1-cocycle</u>.

The function L is said to be a coboundary if there is an $h \in M(A)$ such that

$$(2.7.36)\quad L(\gamma) = h - \beta_{\gamma}(h)$$

This relation is equivalent to

$$(2.7.37)\quad \delta(x) = hx - xh$$

for all $x \in A^{\tau}(\gamma)$, i.e. L is a coboundary if and only if δ is inner.

Furthermore, the derivation δ is bounded if and only if $\|L(\gamma)\|$ is uniformly bounded in γ on $\hat{G}$. If δ is bounded, it follows from 2.7.28 that

$$(2.7.38)\quad \|L(\gamma)\| \leq \|\delta\|$$

for all γ. To prove the converse, assume that A is represented in a faithful G-covariant representation (Such a representation always exist, because the projection $P_0 = \int_G dg\, \alpha_g$ is faithful on A_+, and if ϕ is any state on A^{α}, then $\omega = \phi \circ P_0$ defines an invariant, and thus G-covariant, state on A, see [BR 1], Corollary 2.3.17). We define a transformation T_{γ} of $C = A'' \cap (A^{\alpha})'$ by

$$(2.7.39)\quad T_{\gamma}(h) = L(\gamma) + \beta_{\gamma}(h)$$

The cocycle property 2.7.33 of $L(\gamma)$ implies that T defines an action of G, i.e.

$$(2.7.40)\quad T_{\gamma_1+\gamma_2} = T_{\gamma_1}T_{\gamma_2}$$

for all $\gamma_1, \gamma_2 \in \hat{G}$. As $\|L(\gamma)\|$ is uniformly bounded, the convex closure K of the $L(\gamma)$'s is bounded and thus compact. As $T_{\gamma}(L(\xi)) = L(\gamma + \xi)$, K is T-invariant, i.e. $\gamma \to T_{\gamma}$ is a group of affine maps of the convex compact set K. The Markov-Kakutani fixed point theorem, [Gre 1], implies that K contains a fixed point h for this action. But then $L(\gamma) = h - \beta_{\gamma}(h)$, and hence δ is implemented by h, which means that δ is bounded.

§2.7.2 An example: The circle group and condition Γ

In this section we will consider a simple class of examples where a complete classification on non-commutative vector fields in the sense of §2.1 can be carried out. These examples follows by elaborating example 2.7.8, and are mostly contained in section 3 and 4 of [BGJ 1].

Assume that α is an action of the circle group $\mathbb{T}$ on a C*-algebra $\mathcal{A}$ satisfying the condition

$$(2.7.41) \qquad \overline{A^{\alpha}(n)\, A^{\alpha}(n)^{*}}^{\|\ \|} = A^{\alpha}$$

for all $n \in \hat{\mathbb{T}} = \mathbb{Z}$ (This is called condition Γ in [BGJ 1]). If $\delta \in \mathrm{Der}(\,A_F^{\alpha}\,, \mathcal{A}\,)$ is a derivation with $\delta|\,A^{\alpha} = 0$, let $n \in \mathbb{Z} \mapsto L(n) \in M(\,\mathcal{A}\,) \cap (\,A^{\alpha})'$ be the cocycle associated to α by Example 2.7.8. Then $L(n)$ is uniquely determined by $L(1) = L$ by the formula

$$(2.7.42) \qquad L(n) = \begin{cases} L + \beta(L) + \ldots + \beta^{n-1}(L) & \text{if } n \geq 1 \\ -\beta^{-1}(L) - \beta^{-2}(L) - \ldots - \beta^{n}(L) & \text{if } n \leq -1 \end{cases}$$

which follows from the cocycle relation 2.7.33. Here $\beta = \beta_1$. Conversely, if $L \in M(\,\mathcal{A}\,) \cap (\,A^{\alpha})'$, and one define $L(n)$ by 2.7.42, then $L(n)$ is a cocycle, and if in addition $L = -L^{*}$, then the condition

$$L(-n) = \beta^{-n}(L(n)^{*})$$

is fulfilled.

Hence, there is a one-one linear correspondence between derivations $\delta \in \mathrm{Der}(\,A_F^{\alpha}\,, \mathcal{A}\,)$ satisfying $\delta|_{A^{\alpha}} = 0$ and skew-adjoint operators $L \in M(\,\mathcal{A}\,) \cap (\,A^{\alpha})'$. It follows from 2.7.36 that δ is inner if and only if there exists an element $h \in M(\,\mathcal{A}\,) \cap (\,A^{\alpha})'$ (which can be taken to be skew-adjoint) such that

$$(2.7.43) \qquad L = h - \beta(h)$$

i.e. if and only if

$$(2.7.44) \qquad L \in \mathrm{Range}\ (1 - \beta)$$

Now, assume in addition to 2.7.41 that

(2.7.45) $M(A) \cap (A^\alpha)' \subseteq M(A^\alpha)$.

This condition is meaningful, since $M(A^\alpha) \subseteq M(A)$ by Lemma 2.5.2, and the condition in fact means

$$M(A) \cap (A^\alpha) = \text{centre}\,(M(A^\alpha))$$

In this case L is automatically α-invariant, which means that derivations $\delta \in \mathrm{Der}(A_F^\alpha, A)$ with $\delta|_{A^\alpha} = 0$ automatically commute with α, and hence are pregenerators by Theorem 2.6.1. If in this case δ is approximately inner,

$$\delta(x) = \lim_{n\to\infty} [h_n , x]$$

for $x \in A_F^\alpha$, then it follows from α-invariance of δ that we may replace h_n by $\int_G dg\alpha_g(h_n)$, and hence δ is approximately inner with $h_n \in M(A^\alpha)$. If we now in addition to 2.7.41 and 2.7.45 assume

(2.7.46) A^α is abelian,

then the α-invariant h_n is contained in $(A^\alpha)'$, i.e. $\mathrm{ad}\,(h_n)|_{A^\alpha} = 0$, and thus $L \in \overline{(1-\beta)(M(A^\alpha))}$, where the closure is in the strict topology on $M(A^\alpha)$, i.e. the topology determined by the semi-norms $x \mapsto \|xy\|$, $y \in A^\alpha$. Thus:

<u>Proposition 2.7.9</u> (Bratteli - Goodman - Jørgensen)

Assume that α is an action of the circle group $\mathbb{T}$ on a C*-algebra A satisfying 2.7.41, 45 and 46:

$$\overline{A^\alpha(n)\, A^\alpha(n)^*}^{\|\ \|} = A^\alpha$$

$$M(A) \cap (A^\alpha)' \subseteq M(A^\alpha)$$

$$A^\alpha \subseteq (A^\alpha)'$$

Then there is a one-one linear correspondence between derivations $\delta \in \mathrm{Der}\,(A_F^\alpha, A)$ such that $\delta|_{A^\alpha} = 0$ and skew-adjoint operators $L \in M(A^\alpha)$, given by

(2.7.47) $\delta(x) = Lx$

for $x \in A^\alpha(1)$. The derivation δ is inner if and only if

(2.7.48) $L \in \mathrm{Range}\,(1-\beta) = (1-\beta)(M(A^\alpha))$

where $\beta = \beta_1$ and β_1 the automorphism of $M(A^\alpha)$ defined in Lemma 2.7.2. The derivation δ is approximately inner if and only if

(2.7.49) $L \in \overline{\mathrm{Range}\,(1-\beta)} = \overline{(1-\beta)(M(A^\alpha))}$

where the closure is in the strict topology on $M(A^\alpha)$ (which is the norm topology if A has an identity)

Note that in the situation of Proposition 2.7.9, we have

(2.7.50) $\mathrm{centre}\,(M(A)) \subseteq M(A^\alpha)$

and if δ_o is the generator of the action of $\mathbb{T}$, then the corresponding $L \in \mathrm{centre}\,(M(A^\alpha))$ is

(2.7.51) $L_o = i\mathbb{1}$

Thus, a derivation $\delta \in \mathrm{Der}\,(A_F^\alpha, A)$ has the form

(2.7.52) $\delta = a\delta_o$

with $a = a^* \in \mathrm{centre}\,(M(A))$ if and only if

(2.7.53) $L = ia$

It follows that a general derivation $\delta \in \mathrm{Der}\,(A_F^\alpha, A)$ has a decomposition

(2.7.54) $\delta = a\delta_o + \tilde{\delta}$

where $a \in \mathrm{centre}\,(M(A))$ and $\tilde{\delta}$ is approximately inner, if and only if

(2.7.55) $M(A^\alpha) = \text{centre } (M(A)) + \overline{\text{Range } (1-\beta)}$

and the decomposition is unique if and only if the above decomposition is direct. Note that by definition of β we have

(2.7.56) $\text{centre } (M(A)) = M(A^\alpha)^\beta$

since $\text{centre } (M(A)) \subseteq M(A^\alpha)$. Thus the decomposition 2.7.55 can be written

(2.7.57) $M(A^\alpha) = M(A^\alpha)^\beta + \overline{(1-\beta)(M(A^\alpha))}$

We will now construct a class of examples where 2.7.41, 45 and 46 hold, but nevertheless the decomposition 2.7.57 is not valid. These examples show that the answer of Problem 2.1.3 may be negative.

Example 2.7.10

Let Ω be a compact Hausdorff space and S a homeomorphism of Ω such that the corresponding action of $\mathbf{Z}$ on Ω is free in the sense of Definition 2.4.14. There is an action β of $\mathbf{Z}$ on $B = C(\Omega)$ given as

(2.7.58) $(\beta f)(\omega) = f(S\omega)$

Let $A = B \times_\beta \mathbf{Z}$ be the crossed product of B by this action, [Ped 1], 7.6.5, and let α be the dual action of $\hat{\mathbf{Z}} = \mathbb{T}$ on A, [Ped 1], Proposition 7.8.3. In our case, the crossed product is the closure of the *-algebra of polynomials of the form

(2.7.59) $\sum_n a_n u^n$

where $a_n \in B$, u is a unitary and the sum is finite, with the multiplication law

(2.7.60) $au^n bu^m = au^n bu^{-n} u^{n+m}$

$= a\alpha^n(b) u^{n+m}$

and involution law

$$(2.7.61)\quad (au^n)^* = u^{-n}a^* = u^{-n}a^*u^nu^{-n} = \alpha^{-n}(a^*)u^{-n} ,$$

and the closure is in a certain canonical norm, see [Ped 1]. The dual action of $\mathbb{T} = e^{i\mathbb{R}}$ on A is given by

$$(2.7.62)\quad \alpha_t(au^n) = e^{itn}au^n$$

Any element $x \in A$ has a formal Fourier expansion

$$(2.7.63)\quad x = \sum_{n=-\infty}^{\infty} a_n u^n$$

where

$$(2.7.64)\quad a_n u^n = P_n(x) = \frac{1}{2\pi}\int_0^{2\pi} dt\ e^{-int}\,\alpha_t(x) .$$

(We do not need to worry about the sense in which this series converges for the moment) In particular, the fixed point algebra for the action is

$$(2.7.65)\quad A^{\alpha} = B = C(\Omega)$$

so 2.7.46, $A^{\alpha} \subseteq (A^{\alpha})'$, is fulfilled. Moreover, as each spectral subspace $A^{\alpha}(n)$ contains the unitary u^n , the condition 2.7.41 is fulfilled in the strong form $A^{\alpha}(n)\,A^{\alpha}(n)^* = A^{\alpha}$. Finally, if $x = \sum_{n=-\infty}^{\infty} a_n u^n \in A \cap (A^{\alpha})'$, we have

$$0 = \sum_{n=-\infty}^{\infty} a_n\, u^n\, a - a\, a_n\, u^n = \sum_{n=-\infty}^{\infty} (\alpha^n(a)-a)a_n u^n$$

for all $a \in A^{\alpha}$, and hence

$$(\alpha^n(a) - a)\, a_n = 0$$

for all n by uniqueness of the Fourier decomposition. As the action α is free, we deduce that $a_n = 0$ if $n \neq 0$, and hence $x \in A^{\alpha}$. We have proved 2.7.45, $A \cap (A^{\alpha})' \subseteq A^{\alpha}$.

Next, note that the automorphism $\beta = \beta_1$ on $A^{\alpha} = \mathcal{D}$ defined in Lemma 2.7.2 is nothing but the β defined by 2.7.58, see Remark 2.7.3. Note also that $\overline{\text{Range }(1-\beta)}$ is exactly the annihilator of the set of β-invariant states (i.e. probability measures) on B , this follows from Hahn-Banachs theorem, and the uniqueness of the Jordan decomposition of hermitian functionals (this uniqueness implies that the positive and negative part of a β-invariant functional is still β-invariant).

Now, note that if the algebra $\mathcal{B}$ is $\mathbf{Z}$-simple in the sense of definition 2.6.3, i.e. any S-orbit in Ω is dense in Ω, then A is simple, [EH 1], Corollary 5.16. But if the action of S on Ω is not uniquely ergodic, i.e. if there exists more than one S-invariant probability measure, then since the codimension of $\overline{\text{Range }(1-\beta)}$ in $\mathcal{B}$ is equal to the affine dimension of the compact convex set of invariant probability measures plus one, it follows that the decomposition 2.7.57 is not valid. Thus in this case there exists derivations $\delta \in \text{Der}(A_F^\alpha, A)$ which does not have the decomposition

$$\delta = \lambda\delta_0 + \tilde{\delta}$$

where λ is a scalar and $\tilde{\delta}$ is approximately inner. In fact, the codimension of the space of derivations in $\text{Der}(A_F^\alpha, A)$ having this decomposition is equal to the affine dimension of the convex set of S-invariant probability measures on Ω.

An actual example of a minimal, free, non uniquely ergodic action of $\mathbf{Z}$ on $\Omega = \mathbb{T}^2$ can be found on pp. 18-20 of [EH 1]. For more recent examples see [Maz 1], [Vee 1] and references cited therein.

Note that this also gives another example that the condition that $\hat{G}/\Gamma(\alpha)$ is compact cannot be removed from Theorem 2.6.6. A simple calculation shows that $\Gamma(\alpha) = 0$ in our example.

§2.7.3 <u>Abelian compact groups and invariant derivations</u>

In analyzing invariant derivations we will need the following variant of Lemma 2.7.5.

<u>Lemma 2.7.11</u>

Let α be an action of an abelian compact group G on a C^*-algebra A covariantly and faithfully represented on a Hilbert space H. Let δ be a (norm-) closed derivation defined on a dense $*$-subalgebra $D(\delta)$ of A and with values in $M = A''$, such that

(2.7.67) $\quad A^\alpha \subseteq D(\delta)$ and $\delta\big|_{A^\alpha} = 0$

and

(2.7.68) $[\delta, \alpha] = 0$ (See Definition 2.5.9)

It follows that for each $\gamma \in \hat{G}$ there exists a (possibly unbounded) skew-adjoint operator $L(\gamma)$ affiliated with the abelian von Neumann algebra $(M^\alpha \cap (M^\alpha)')E(\gamma)$, where $E(\gamma)$ is the projection onto $[M^\alpha(\gamma)H]$, such that

(2.7.69) $(D(\delta) \cap A^\alpha(\gamma))H \subseteq D(L(\gamma))$

and

(2.7.70) $L(\gamma)x\xi = \delta(x)\xi$

for each $x \in D(\delta) \cap A^\alpha(\gamma)$, $\xi \in H$.

Furthermore, $A^\alpha(\gamma)\, A^\alpha(\gamma)^*$ has an approximate identity of the form

(2.7.71) $e_\tau = \sum_i a_i^\tau a_i^{\tau *}$

where $a_i^\tau \in D(\delta) \cap A^\alpha(\gamma)$, $(D(\delta) \cap A^\alpha(\delta))H$ is a core for $L(\gamma)$, and $L(\gamma)$ is defined on this core by

(2.7.72) $L(\gamma)x\xi = \lim_{\tau\to\infty} (\sum_i \delta(a_i^\tau)a_i^{\tau *})x\xi$

for $x \in D(\delta) \cap A^\alpha(\gamma)$, $\xi \in H$.

Proof

This Lemma is contained in the proof of Theorem 5.1 in [BJ 1]. First note that $D(\delta) \cap A^\alpha(\gamma)$ is dense in $A^\alpha(\gamma)$, $\delta(D(\delta) \cap A^\alpha(\gamma)) \subseteq M^\alpha(\gamma)$ and $D(\delta) \cap A_F^\alpha$ is a core for δ by the reasoning in Lemma 2.5.8. Then the existence of an approximate identity of the given form follows from the proof of Lemma 2.5.1 (Choose the finite sets $\{x_i\}$ to be contained in $D(\delta) \cap A^\alpha(\gamma)$ instead of just $A^\alpha(\gamma)$).

The existence of the limit 2.7.72 now follows as in 2.7.75. Putting

$$(2.7.73)\quad L_\tau(\gamma) = \sum_i \delta(a_i^\tau) a_i^{\tau *} ,$$

each $L_\tau(\gamma)$ is skew-adjoint by the identity

$$(2.7.74)\quad 0 = \delta(\sum_i a_i^\tau a_i^{\tau *}) = \sum_i \delta(a_i^\tau) a_i^{\tau *} + \sum_i a_i^\tau \delta(a_i^{\tau *}) = L_\tau(\gamma) + L_\tau(\gamma)^*$$

and hence the limit $L(\gamma)$ defined by 2.7.72 is skew-symmetric and thus closable. Denoting the closure again by $L(\gamma)$, it follows from

$$(2.7.75)\quad L_\tau(\gamma) \in A^\alpha(\gamma)\, A^\alpha(\gamma)^*$$

that $L(\gamma)$ is affiliated to $M^\alpha E(\gamma)$, while $L(\gamma)$ is affiliated to $(M^\alpha)'E(\gamma)$ since $\delta(ax) = a\delta(x)$ for $a \in A^\alpha$, $x \in D(\delta) \cap A^\alpha(\delta)$. Thus $L(\gamma)$ is affiliated to the abelian von Neumann algebra $(M^\alpha \cap (M^\alpha)')E(\gamma)$, and since this algebra is finite and $L(\gamma)$ is skew-symmetric, $L(\gamma)$ is actually skew-adjoint, [Seg 1].

§2.7.4 General compact groups

Let G be a compact group and α an action of G on a C*-algebra A. In analyzing derivations δ defined on A_F^α with $\delta|_{A^\alpha} = 0$, some new difficulties occurs. If $\gamma \in \hat{G}$ and $x \in A_n^\alpha(\gamma)$ (see Definition 2.2.12), then it follows from 2.2.42 that

$$(2.7.76)\quad \alpha_g(xx^*) = xx^*$$

for all $g \in G$, but

$$(2.7.77)\quad \alpha_g(x^*x) = (\mathbb{1}_A \otimes \gamma(g)^*)(x^*x)(\mathbb{1}_A \otimes \gamma(g))$$

Thus $xx^* \in A^\alpha \otimes M_n$, but $x^*x \notin A^\alpha \otimes M_{d(\gamma)}$ in general unless the representation γ is one-dimensional. In fact, if one defines

$$M_\gamma^\alpha = \{x \in A \otimes M_{d(\gamma)} | \alpha_g(x) = (\mathbb{1} \otimes \gamma(g))^*\, x\, (\mathbb{1} \otimes \gamma(g))$$
$$\text{for all } g \in G\}$$

then $x^*x \in M_\gamma^\alpha$, and since

$$A_n^\alpha(\gamma) M_\gamma^\alpha \subseteq A_n^\alpha(\gamma)$$

the linear span $A_n^\alpha(\gamma)^* A_n^\alpha(\gamma)$ is an ideal in the C*-algebra M_γ^α.

Thus computations like 2.7.23 break down when G is not abelian. Another difficulty arises from the fact that $A_n^\alpha(\gamma)$ is not α-invariant; if $x \in A_n^\alpha(\gamma)$ is nonzero it follows from 2.2.42 that $\alpha_g(x) \in A_n^\alpha(\gamma)$ if and only if $\gamma(g)$ is a scalar multiple of the identity.

Still another problem is that the analogue of the isomorphism β_γ of Lemma 2.7.2 is an isomorphism β_γ from the relative commutant of $A_1^\alpha(\gamma)^* A_1^\alpha(\gamma)$ in the multiplier algebra of the norm closure of the subalgebra $A_1^\alpha(\gamma)^* A A_1^\alpha(\gamma)$ of $A \otimes M_{d(\gamma)}$ onto the relative commutant of $A_1^\alpha(\gamma) A_1^\alpha(\gamma)^*$ in the multiplier algebra of the norm closure of the subalgebra $A_1^\alpha(\gamma) A A_1^\alpha(\gamma)^*$ of A, given by

$$\beta_\gamma(a)x = xa$$

for all $x \in A_1^\alpha(\gamma)$, $a \in (A_1^\alpha(\gamma)^* A_1^\alpha(\gamma))' \cap M\left(\overline{A_1^\alpha(\gamma)^* A A_1^\alpha(\gamma)}^{\|\ \|}\right)$. This β_γ does not seem particularly useful for our purposes.

Thus we do not have a completely satisfactory analogue of Lemma 2.7.5 at present, but under somewhat special circumstances and when δ is a derivation we can define an analogue of the partial cocycle $\gamma \to R(\gamma)$:

Lemma 2.7.12 (Bratteli-Goodman)

Let α be a σ-weakly continuous action of a compact group G on a von Neumann algebra M on a Hilbert space H such that

$$(2.7.78)\quad (M^\alpha)' \cap M \subseteq M^\alpha$$

Let δ be a derivation from M_F^α into M such that

$$(2.7.79)\quad \delta\big|_{M^\alpha} = 0.$$

If $[x_{ij}] \in M_n^\alpha(\gamma)$, $\gamma \in \hat{G}$, $n = 1,2,\ldots$, see Definition 2.2.12, define

$$(2.7.80)\quad \delta([x_{ij}]) = [\delta(x_{ij})].$$

For each $\gamma \in \hat{G}$ of dimension $d(\gamma)$, there exists an skewsymmetric operator $R(\gamma)$ affiliated with $(M \cap (M^\alpha)') \otimes M_{d(\gamma)}$ such that

(2.7.81) $\delta(X) = XR(\gamma)$

for all $X \in M_1^\alpha(\gamma)$.

Proof

This is Observation 4 in the proof of Theorem 2.5 in [BG 1]. (The proof in [BG 1] has a gap.)

Let $\mathbf{F}$ be the union over $n \in \mathbf{N}$ of the partial isometries in $M_n^\alpha(\gamma)$, and give $\mathbf{F}$ the following partial order: $u \leq_{\mathbf{F}} v$ if $u^*u \leq v^*v$ in $M \otimes M_d$. Observe that $\mathbf{F}$ is directed since if $u,v \in \mathbf{F}$ and w is the partial isometry part of $\begin{bmatrix} u \\ v \end{bmatrix}$, then w^*w dominates both u^*u and v^*v. For each $u \in \mathbf{F}$, define $R(u) = u^*\delta(u) \in M \otimes M_d$. If $x \in M_1^\alpha(\gamma)$ has the polar decomposition $x = (xx^*)^{\frac{1}{2}}u$ and if $v \geq_{\mathbf{F}} u$, then $x = xv^*v$ and

$$\delta(x) = xv^*\delta(v) = xR(v),$$

since xv^* is a matrix over M^α. Hence if $v \geq_{\mathbf{F}} u$, then

(2.7.82) $u^*uR(u) = u^*\delta(u) = u^*uR(v)$.

We define an operator $R(\gamma)$ on $\bigcup_{u \in \mathbf{F}} u^*u \cdot H^d \equiv H(\gamma)$ by putting $R(\gamma)\xi = -R(u)^*\xi$ for $\xi \in u^*u \cdot H^d$. Note that $H(\gamma)$ is a vector space since $\mathbf{F}$ is directed, and 2.7.82 implies that $R(\gamma)$ is well defined on $H(\gamma)$. Let $P(\gamma)$ be the range projection on $\overline{H(\gamma)}$. We claim that $P(\gamma)R(\gamma)P(\gamma)$ is skewsymmetric on $H(\gamma)$. Given $\xi,\eta \in H(\gamma)$, choose $u \in \mathbf{F}$ such that $u^*u\xi = \xi$ and $u^*u\eta = \eta$. Then

$$\begin{aligned}
(\eta, R(\gamma)\xi) &= -(\eta, R(u)^*u^*u\xi) \\
&= -(u^*u\eta, \delta(u)^*u\xi) \\
&= -(u\eta, u\delta(u^*)u\xi) \\
&= -(u\eta, [\delta(uu^*) - \delta(u)u^*]u\xi) \qquad (2.7.83) \\
&= (\delta(u)^*u\eta, u^*u\xi) \\
&= (R(u)^*\eta, \xi) \\
&= (-R(\gamma)\eta, \xi)
\end{aligned}$$

Thus $P(\gamma)R(\gamma)P(\gamma)$ is skew-symmetric. We extend $R(\gamma)$ to a densely defined operator on $H^{d(\gamma)}$ by putting

$$R(\gamma)(\mathbf{1} - P(\gamma)) = 0 \tag{2.7.84}$$

Next we show that $R(\gamma)$ is affiliated with $(M^{\alpha})' \otimes M_d$. If $v \in M^{\alpha}$ is unitary and $u \in \mathbf{F}$, then $u(v \otimes \mathbf{1}_d) \in \mathbf{F}$, so $(v^* \otimes \mathbf{1}_d)H(\gamma) = H(\gamma)$. Take $\xi \in H(\gamma)$ and choose $u \in \mathbf{F}$ such that $u^*u\xi = \xi$. Then

$$\begin{aligned} &(v \otimes \mathbf{1}_d)R(\gamma)(v^* \otimes \mathbf{1}_d)\xi \\ &= (v \otimes \mathbf{1}_d)R(\gamma)(v^* \otimes \mathbf{1}_d)u^*u\xi \\ &= -(v \otimes \mathbf{1}_d)\delta[(v^* \otimes \mathbf{1}_d)u^*]u\xi \\ &= -\delta(u)^*u\xi = -R(u)^*\xi = R(\gamma)\xi. \end{aligned} \tag{2.7.85}$$

Since $P(\gamma) \in (M^{\alpha} \otimes \mathbf{1}_d)' = (M^{\alpha})' \otimes M_d$ it follows that $R(\gamma)$ is affiliated with $(M^{\alpha})' \otimes M_d$. Also, $R(\gamma)$ is affiliated with $M \otimes M_d$ since $R(\gamma)|_{H(\gamma)}$ is the limit of operators $-R(u)^* = -\delta(u^*)u \in (M \otimes M_d)P(\gamma)$

Finally, for $u \in \mathbf{F}$,

$$\begin{aligned} u^*uR(\gamma) &= u^*uP(\gamma)R(\gamma)P(\gamma) \\ &= (-P(\gamma)R(\gamma)P(\gamma)u^*u)^* \\ &= (-P(\gamma)R(\gamma)u^*u)^* \\ &= (P(\gamma)R(u)^*u^*u)^* \\ &= u^*uR(u)P(\gamma) \end{aligned} \tag{2.7.86}$$

Hence if $x \in M_1^{\alpha}(\gamma)$, and $x = |x|u$ is its polar decomposition, then

$$u \in M_1^{\alpha}(\gamma), \quad |x| \in M^{\alpha}$$

by Lemma 2.7.7 and hence

$$\begin{aligned} xR(\gamma) &= xu^*uR(\gamma) = xu^*uR(u)P(\gamma) \\ &= xR(u)P(\gamma) \\ &= \delta(x)P(\gamma) \end{aligned} \tag{2.7.87}$$

for $x \in M_1^\alpha(\gamma)$. At this stage in the proof we need the auxilliary condition 2.7.78. This condition implies that $P(\gamma)$, which is the range projection of $[M_1^\alpha(\gamma)^* M_1^\alpha(\gamma) H^d]$, is contained in $M^\alpha \otimes M_d \supseteq ((M^\alpha)' \cap M) \otimes M_d$. As $\delta\big|_{M^\alpha \otimes M_d} = 0$ it follows from the derivation property that

$$(2.7.88) \quad \delta(xP(\gamma)) = \delta(x)P(\gamma)$$

for any $n \times d$ matrix $x = [x_{ij}]$ with $x_{ij} \in M_F^\alpha$. Thus, from 2.7.87

$$(2.7.89) \quad \begin{aligned} xR(\gamma) &= \delta(xP(\gamma)) \\ &= \delta(x) \end{aligned}$$

for $x \in M_1^\alpha(\gamma)$, which is 2.7.81. Also, as $R(\gamma)$ is the limit of the operators $-R(u)^* = -\delta(u^*)u$ on $H(\gamma)$ and

$$(2.7.90) \quad \begin{aligned} P(\gamma)\delta(u^*)u &= \delta(P(\gamma)u^*)u \\ &= \delta(u^*)u \end{aligned}$$

from 2.7.88, it follows that

$$(2.7.91) \quad P(\gamma)R(\gamma) = R(\gamma) = R(\gamma)P(\gamma)$$

Thus $P(\gamma)$ commute with $R(\gamma)$, and $R(\gamma)$ is actually skew-adjoint. This ends the proof of Lemma 2.7.12.

<u>Problem 2.7.13</u>

Is Lemma 2.7.12 still true if condition 2.7.78 is removed? Note that this condition could be replaced by the weaker condition that the range projection of $[M_1^\alpha(\gamma)^* M_1^\alpha(\gamma) H^{d(\gamma)}]$ is contained in $M^\alpha \otimes M_{d(\gamma)}$ for each $\gamma \in \hat{G}$.

§2.8. Invariant derivations

In this section we will study closed derivations invariant under a group of automorphisms, but without any assumptions on the domain except for density.

§2.8.1. G compact, abelian; $A^{\alpha} \subseteq D(\delta)$

Theorem 2.8.1 (Bratteli-Jørgensen-Kishimoto)

Let G be a compact, abelian group and α an action of G on a C*-algebra A. Let δ be a closed derivation of A such that

(2.8.1) $[\delta, \alpha] = 0$ (See Definition 2.5.9),

and

(2.8.2) $A^{\alpha} \subseteq D(\delta)$,

where A^{α} is the fixed point algebra for the action α.

It follows that δ is a generator.

Remark 2.8.2

This theorem fails if 2.8.2 is replaced by the condition: $\delta|_{A^{\alpha}}$ is a generator, see Example 2.8.5.

The theorem also fails if the condition that G is compact is removed, see Examples 2.8.6 and 2.8.7.

It is unknown if the theorem holds in general if the condition that G is abelian is removed, but it still holds when G is non-abelian if A is abelian, see Theorem 2.4.32, or if $\delta|_{A^{\alpha}} = 0$ and A is type I, see Theorem 2.8.16.

In the case that $\delta|_{A^{\alpha}} = 0$ (or inner) the theorem was shown independently in [BJ 1], Theorem 5.1, and by Kishimoto, see [Iku 1], Appendix. The present generalization was proved in [BK 1]. It was announced in a postscript to [Pel 2].

Proof

Assume that A is faithfully and covariantly represented on a Hilbert space H, let $M = A''$, and let α also denote the extension of α to M. We first prove that δ is σ-weakly closable, and its σ-weak closure $\overline{\delta}$ generates a σ-weakly continuous one-parameter group of automorphisms of M.

We have $\delta(A^\alpha) \subseteq A^\alpha$ by Proposition 2.5.8, and it follows from Sakai-Kadison's derivation theorem that there exists a $h = -h^* \in M^\alpha = A^{\alpha''}$ such that

(2.8.3) $\delta(x) = hx - xh$

for all $x \in A^\alpha$, see Theorem 2.5.11. Define derivations $\tilde{\delta}, \delta_0 : D(\delta) \to M$ by

(2.8.4) $\tilde{\delta}(x) = hx - xh \ , \quad \delta_0(x) = \delta(x) - \tilde{\delta}(x)$

for $x \in D(\delta)$. As $h \in M^\alpha$, it follows that

(2.8.5) $[\tilde{\delta}, \alpha] = 0 \ , \quad [\delta_0, \alpha] = 0 \ .$

By definition of h,

(2.8.6) $\delta_0\big|_{A^\alpha} = 0 \ .$

By Lemma 2.7.11, for each $\gamma \in \hat{G}$ there exists a (possibly unbounded) closed, skew-adjoint operator $L(\gamma)$ affiliated with the abelian von Neumann algebra $(M^\alpha \cap (M^\alpha)')E(\gamma)$, where $E(\gamma)$ is the projection onto $[M^\alpha(\gamma)H]$, such that

(2.8.7) $\delta_0(x) = L(\gamma)x$

for each $x \in D(\delta) \cap A^\alpha(\gamma)$. Then $e^{tL(\gamma)}$ is a unitary operator in $(M^\alpha \cap (M^\alpha)')E(\gamma)$ for each $t \in \mathbb{R}$, and hence

(2.8.8) $\tau_t(x) = e^{tL(\gamma)}x \ , \ x \in M^\alpha(\gamma) \ ,$

defines a σ-weakly continuous group of isometries of $M^\alpha(\gamma)$ for each $\gamma \in \hat{G}$. Define a oneparameter group τ_t of M_F^α by

$$(2.8.9)\quad \tau_t(\sum_\gamma x_\gamma) = \sum_\gamma e^{tL(\gamma)}x_\gamma$$

when $x_\gamma \in M^\alpha(\gamma)$.

Now, as every isomorphism between two von Neumann algebras in a standard representation is spatial, it follows that any such isomorphism has a natural extension to closed, unbounded operators affiliated to the algebra. If $\beta_\gamma\colon (M^\alpha \cap (M^\alpha)')E(-\gamma) \to (M^\alpha \cap (M^\alpha)')E(\gamma)$ is the restriction of the autormorphism β_γ defined in Lemma 2.7.2 to the center of M^α, it follows as in Lemma 2.7.5 that

$$(2.8.10)\quad L(\gamma_1+\gamma_2)E(\gamma_1,\gamma_2) = L(\gamma_1)E(\gamma_1,\gamma_2)+\beta_{\gamma_1}(E(-\gamma_1)L(\gamma_2))E(\gamma_1,\gamma_2)$$

see 2.7.17. But all operators in this expression is affiliated to the abelian von Neumann algebra $M^\alpha \cap (M^\alpha)'$, hence after exponentiation and multiplication by $E(\gamma_1,\gamma_2)$:

$$(2.8.11)\quad e^{tL(\gamma_1+\gamma_2)}E(\gamma_1,\gamma_2) = e^{tL(\gamma_1)}\beta_{\gamma_1}(E(-\gamma_1)e^{tL(\gamma_2)})\,E(\gamma_1,\gamma_2)$$

where we have extended the operators $L(\gamma)$ to H by defining $L(\gamma)(\mathbb{1}-E(\gamma)) = 0$. Thus

$$(2.8.12)\quad e^{tL(\gamma_1+\gamma_2)}x_1x_2 = e^{tL(\gamma_1)}\beta_{\gamma_1}(E(-\gamma_1)e^{tL(\gamma_2)})x_1x_2 = e^{tL(\gamma_1)}x_1e^{tL(\gamma_2)}x_2$$

for $x_i \in M^\alpha(\gamma_i)$, $i = 1, 2$ or

$$(2.8.13)\quad \tau_t(x_1x_2) = \tau_t(x_1)\tau_t(x_2)$$

Thus τ is a one-parameter group of automorphisms of M_F^α . But using 2.7.18 in a similar manner, we verify that τ is a group of *-automorphisms. It now follows from Lemma 2.5.4 that τ is a group of isometries.

Now, the projection $P_0 = \int_G dg\,\alpha_g$ from M onto M^α is faithful and normal. If ω is a normal state on M^α, then $\omega\circ P_0$ defines a normal state on M, and as $L(0) = 0$, this state is τ-invariant. Thus the direct sum of the representations defined by all states of M of this form defines a faithful normal representation of M where the group τ on M_F^α is unitary implemented by a strongly continuous unitary representation, see [BR 1], Corollary 2.3.17. As M_F^α is σ-weakly dense in M, it follows that τ extends to a σ-weakly continuous one-parameter group of *-automorphisms of M. (Actually, the above argument gives a proof of the isometric property of τ without using Lemma 2.5.4.) Denote the σ-weak generator of this group by δ_τ, see [BR 1], Definition 3.1.5. We will now argue that δ_τ is nothing but the σ-weak closure of the earlier defined δ_0 on $D(\delta)$.

We first argue that δ_τ extends δ_0. But as $D(\delta) \cap A_F^\alpha$ is a core for δ_0 by 2.5.27, it suffices to show that:

(2.8.14) $D(\delta) \cap A^\alpha(\gamma) \subseteq D(\delta_\tau)$ and $\delta_\tau(x) = L(\gamma)x$.

for $x \in D(\delta) \cap A^\alpha(\gamma)$.

But if $x \in D(\delta) \cap A^\alpha(\gamma)$, it follows from

(2.8.15) $$\tau_t(x) - x = e^{tL(\gamma)}x - x = \int_0^t ds\, e^{sL(\gamma)}L(\gamma)x$$

and the strong continuity of $s \to e^{sL(\gamma)}$ that

(2.8.16) $$\frac{d}{dt}\Big|_{t=0} \tau_t(x) = L(\gamma)x$$

where the derivative is in the σ-weak topology. This shows 2.8.14, i.e. δ_τ extends δ_0, and in particular δ_0 is σ-weakly closable. Let $\overline{\delta_0}$ be the σ-weak closure.

We want to show that $\overline{\delta_0} = \delta_\tau$. To that end we need:

Lemma 2.8.3

The elements $x \in D(\delta) \cap A^\alpha(\gamma)$ such that there exists an $y \in D(\delta) \cap A^\alpha(\gamma)$ with $x = yy^*x$ is a norm dense subset of $A^\alpha(\gamma)$. Furthermore this subset consists of entire analytic elements for $\overline{\delta_0}$.

Proof

Define real, positive continuous functions f_n, g_n by

$$f_n(t) = \begin{cases} 0 & \text{for } 0 \leq t \leq \frac{1}{n}, \\ nt-1 & \text{for } \frac{1}{n} \leq t \leq \frac{2}{n}, \\ 1 & \text{for } \frac{2}{n} \leq t, \end{cases}$$

and

$$g_n(t) = \begin{cases} n^{3/2}\, t & \text{for } 0 \leq t \leq \frac{1}{n}, \\ (t)^{-\frac{1}{2}} & \text{for } \frac{1}{n} \leq t, \end{cases}$$

and if $z \in D(\delta) \cap A^{\alpha}(\gamma)$ is given, define

$$x_n = f_n(zz^*)z, \quad y_n = g_n(zz^*)z.$$

Then, as $zz^* \in A^{\alpha}$, $x_n, y_n \in A^{\alpha}(\gamma) \cap D(\delta)$ and as $g_n(t)^2 t = 1$ for $t \in \mathrm{Supp}(f_n)$, we have

$$x_n = y_n y_n^* x_n$$

But by Lemma 2.7.7, z has the form

$$z = (zz^*)^{\frac{1}{4}} w$$

where $w \in A^{\alpha}(\gamma)$. But as

$$\| f_n(zz^*)(zz^*)^{\frac{1}{4}} - (zz^*)^{\frac{1}{4}} \| \to 0$$

as $n \to \infty$, it follows that

$$x_n = f_n(zz^*)(zz^*)^{\frac{1}{4}} w \to (zz^*)^{\frac{1}{4}} w = z$$

as $n \to \infty$. This proves the density statement of the lemma.

Next we prove that any $x \in D(\delta) \cap A^{\alpha}(\gamma)$ of the form $x = yy^*x$, where $y \in D(\delta) \cap A^{\alpha}(\gamma)$, is entire analytic for $\overline{\delta_0}$. As $y^*x \in A^{\alpha}$, we have

$$\delta_0(x) = \delta_0(yy^*x) = \delta_0(y)y^*x$$

But $\delta_0(y) \in M^{\alpha}(\gamma)$, hence $\delta_0(y)y^* \in M^{\alpha}$. The module property

$$\delta_0(ax) = a\delta_0(x)$$

for $a \in A^\alpha$, $x \in A^\alpha(\gamma)$ immediately implies that $M^\alpha(A^\alpha(\gamma) \cap D(\delta)) \subseteq D(\overline{\delta}_0)$, and the module property extends by closure to $a \in M^\alpha = (A^\alpha)''$. Thus $\delta_0(y)y^*x \in D(\overline{\delta}_0)$ and

$$\overline{\delta}_0(\delta_0(y)y^*x) = \delta_0(y)y^*\delta_0(x) = (\delta_0(y)y^*)^2x$$

Iterating, we deduce that $x \in D(\overline{\delta}_0{}^n)$ and

$$\overline{\delta}_0{}^n(x) = (\delta_0(y)y^*)^n x$$

for $n = 1, 2, \ldots.$ This establishes that x is entire analytic for $\overline{\delta}_0$.

This ends the proof of Lemma 2.8.3.

Lemma 2.8.3 implies that $D(\overline{\delta}_0)$ is a core for the generator δ_τ, see [BR 1], Proposition 3.2.58. Thus

$$\delta_\tau = \overline{\delta}_0$$

But as $\tilde{\delta} = \mathrm{ad}(h)$ is a bounded, σ-weakly continuous derivation of M, this establishes that $\delta = \delta_0 + \tilde{\delta}$ is σ-weakly closable with closure

$$\overline{\delta} = \delta_\tau + \tilde{\delta}\ , \quad D(\overline{\delta}) = D(\delta_\tau)\ ,$$

and this closure is a generator, see [BR 1], Theorem 3.1.33.

This establishes that δ is σ-weakly closable in all faithfull G-covariant representations of A, and the σ-weak closure $\overline{\delta}$ is a generator of a σ-weakly continuous oneparameter group of *-automorphisms of A''. Thus Theorem 2.8.1 follows from the next lemma.

<u>Lemma 2.8.4</u>

Let G be a compact, abelian group and α an action of G on a C*-algebra A. Let δ be a closed derivation of A such that

$$[\delta, \alpha] = 0$$

and the σ-weak closure $\overline{\delta}$ of δ exists and is a generator of a σ-weakly continuous oneparameter group of *-automorphisms of A'' in each faithful G-covariant representation of A.

It follows that δ is a generator.

Proof

By Theorem 1.5.2 and Definition 1.4.6 we have to show

(2.8.17) $\|(1 + \lambda\delta)(x)\| \geq \|x\|$

for all $x \in D(\delta)$ and all real λ, and

(2.8.18) $\overline{(1 + \lambda\delta)(D(\delta))}^{\|\ \|} = A$

for all real λ.

But if π is any faithful G-covariant representation, then

$$\|\pi((1 + \lambda\delta)(x))\| \geq \|\pi(x)\|$$

by the generator assumption, see [BR 1], Theorem 3.1.10. As π is an isometry, [BR 1], Corollary 2.2.6, the estimate 2.8.17 follows.

Next, assume ad absurdum that 2.8.18 is false. Then there exists a non-zero functional $\eta \in A^*$ such that

$$\eta((1 + \lambda\delta)(x)) = 0$$

for all $x \in D(\delta)$. But as $[\delta, \alpha] = 0$, it follows that

$$\eta(\alpha_g((1 + \delta)(x))) = \eta((1 + \delta)(\alpha_g(x))) = 0$$

for all $x \in D(\delta)$. Thus if

$$\eta_\gamma = \int_G dg\ \overline{\gamma(g)}\ \eta\circ\alpha_g$$

denotes the Fourier coefficients for η, where $\gamma \in \hat{G}$, then

$$\eta_\gamma \circ (1 + \lambda\delta) = 0$$

for all $\gamma \in \hat{G}$. But as $\eta \neq 0$, $\eta_\gamma \neq 0$ for some γ. Thus we may assume that η has the transformation law

$$\eta \circ \alpha_g = \gamma(g)\eta$$

for some $\gamma \in \hat{G}$ and all $g \in G$. Let

$$\eta = |\eta|u$$

be the polar decomposition of η in the bidual A^{**} of A, where $|\eta|$ is a positive functional and u a partial isometry in A^{**} with range projection = support of $|\eta|$, see [Sak 6], Theorem 1.14.4. But

$$\eta \circ \alpha_g = \gamma(g)\eta = |\eta|(\gamma(g)u) ,$$

and as this is the polar decomposition of $\eta \circ \alpha_g$ it follows from uniqueness of this decompostition that

$$|\eta| \circ \alpha_g = |\eta \circ \alpha_g| = |\eta| ,$$

i.e. $|\eta|$ is α-invariant, and thus $|\eta|$ is a vector state in a faithful G-covariant representation. But then $\eta = |\eta|u$ is a vector functional in the same representation. But as

$$\eta((1 + \lambda\bar{\delta})(D(\delta))) = 0,$$

it follows that

$$\eta((1 + \lambda\bar{\delta})(D(\bar{\delta})) = 0$$

where $\bar{\delta}$ is the σ-weak closure of δ in the representation. But since $\bar{\delta}$ is a generator, $(1 + \lambda\bar{\delta})(D(\bar{\delta})) = A''$ and hence $\eta = 0$. This contradiction establishes Lemma 2.8.4 and hence Theorem 2.8.1.

Example 2.8.5 ([BJ 1], Example 6.1)

The conclusion of Theorem 2.8.1 may fail if the condition "$A^\alpha \subseteq D(\delta)$" is replaced by the condition "$\delta|_{A^\alpha}$ is a generator". If $A = C_0(\Omega)$ is abelian this can be understood as follows: in this case $A^\alpha \subseteq D(\delta)$ means that $\delta|_{A^\alpha} = 0$, and hence δ is tangential to the

G-orbits in Ω in a sense made precise in Theorem 2.4.32. If we only require that $\delta|_{A^\alpha}$ is a generator, δ can be transverse to the G-orbits, however. A concrete example can be constructed by considering the two-torus $\mathbb{T}^2$ with 2π-periodic coordinates (θ, ϕ). Let Ω be the compact Hausdorff space obtained from $\mathbb{T}^2$ by compressing $\{(\theta, \phi) | \pi \leqslant \phi \leqslant 2\pi\}$ to a ϕ-interval $[\pi, 2\pi]$, i.e. the points (θ_1, ϕ) and (θ_2, ϕ) are identified when $\pi \leqslant \phi \leqslant 2\pi$. The space Ω may be embedded in $\mathbb{R}^3$ as a sausage on a string:

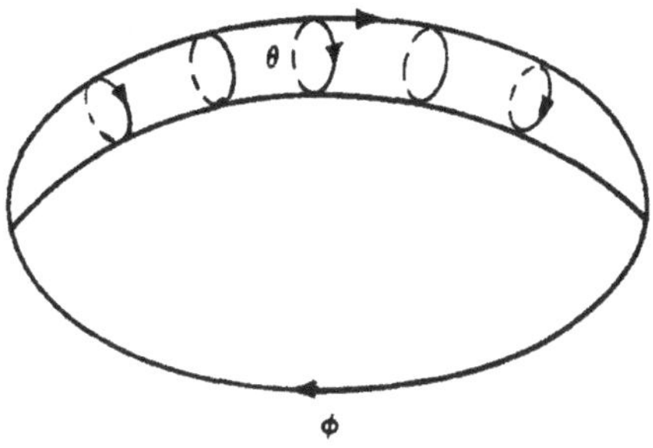

Let $A = C(\Omega)$, $\delta = \frac{\partial}{\partial\phi}$, $G = \mathbb{T}$, and let G act on Ω by translation in the θ-variable. This action has circular orbits when $0 < \phi < \pi$ and point orbits when $\pi < \phi < 2\pi$. Clearly δ commutes with this action α, A^α identifies with $C(\mathbb{T})$ in such a way that $\delta|_{A^\alpha}$ is the generator of rotations on $\mathbb{T}$, but δ is not a generator. The last is clear from geometric considerations: δ cannot generate a flow locally at the two singular points $\phi = 0$, π.

<u>Example 2.8.6</u> (Takesaki, see [BJ 1], Example 6.5)

In the case that G is not compact, the condition $A^\alpha \subseteq D(\delta)$ is not sufficiently severe to ensure that δ is a generator, see Section 2.11. -- for a more detailed discussion. An example of this is the even CAR algebra A over the Hilbert space $H = L^2(\mathbb{R}) \otimes K$, where K is an infinite dimensional Hilbert space, [BR2], Theorem 5.2.5. Let $t \in \mathbb{R} \to U_t$ be the translation group in the first variable (or any unitary group on $L^2(\mathbb{R})$ with absolutely continuous spectrum), and let H be a symmetric opperator on K with deficiency indices $(0, 1)$. Let $t \to \alpha_t$ be the group of quasi-free $*$-automorphisms defined by $U_t \otimes 1$ and δ the quasifree derivation defined by

$1 \otimes H$, [Bra 4] .
Then α is asymptotically abelian in norm, [BR 2], Example 5.2.21, and as A is simple this means that α is ergodic in the sense that $A^{\alpha} = \mathbb{C}\,\mathbb{1}$, hence the condition $\delta|_{A^{\alpha}} = 0$ is trivially fullfilled while $[\delta, \alpha] = 0$ follows from $[1 \otimes H, U \otimes 1] = 0$. As $1 \otimes H$ generates a one-parameter semigroup of isometries which are not unitaries, δ generates a one-parameter semigroup of endomorphisms which are not automorphisms, thus δ is not a generator of a group of automorphisms, and does not even extend to such a generator on A .

Example 2.8.7 ([BDR 1], Example 2.4, [Bat 4], Example 6.5)

This is an example of an action of $G = \mathbb{R}$ on an abelian C*-algebra $A = C(\Omega)$ where the G-orbits on Ω is closed, such that the condition $\delta|_{A^{\alpha}}$ forces δ to be tangential to the G-orbits in the sense

$$\mathrm{supp}(\delta(f)) \subseteq \mathrm{supp}(\delta_0(f))$$

for $f \in D(\delta) \cap D(\delta_0)$, where δ_0 is the generator of the $\mathbb{R}$-action, but where δ nevertheless is not a generator.

Let $\Omega = [0, 1] \times [0, 1]$ be the unit square. Further let a be a continuous function over $[0, 1]$ such that $a(x) > 0$ for $0 < x < 1$, $a(0) = 0 = a(1)$, and

$$\int_0^1 dx \frac{1}{a(x)} < +\infty .$$

Next define functions g and h on Ω as follows:

$$g(x, y) = \begin{cases} \inf\{a(x), x/y\} & \text{if } 0 < x < 1/2 \\ \inf\{a(x), (1-x)/y\} & \text{if } 1/2 < x < 1 \end{cases}$$

for $y \neq 0$, and

$$g(x, 0) = a(x) ;$$

$$h(x, y) = yg(x, y) .$$

Now g and h are continuous functions on Ω and hence

$$(\delta_0 f)(x, y) = h(x, y) \frac{\partial f}{\partial x} (x, y),$$

$$(\delta f)(x, y) = g(x, y) \frac{\partial f}{\partial x} (x, y)$$

are well defined, norm closed derivations of $C(\Omega)$. Moreover since

$$\int_0^{\varepsilon} dx \frac{1}{h(x,y)} = +\infty = \int_{1-\varepsilon}^{1} dx \frac{1}{h(x,y)}$$

for all y, δ_0 is the generator of a flow in the x-direction by the initial remarks in Section 2.4.2. (These latter conditions ensure that a point in the interior $\langle 0, 1\rangle \times [0, 1]$ of Ω never hits the boundary.) But since

$$\int_0^{\varepsilon} dx \frac{1}{g(x,0)} < +\infty , \qquad \int_{1-\varepsilon}^{1} dx \frac{1}{g(x,0)} < +\infty$$

δ is not a generator (because the point $(x, 0)$ hits the boundary in a finite time). Now g is proportional to h on each orbit and hence δ commutes with the flow defined by δ_0 . Moreover as g and h have the same closed support one easily verifies that $\mathrm{supp}(\delta(f)) \subseteq \mathrm{supp}(\delta_0(f))$ for all $f \in D(\delta) \cap D(\delta_0)$. In particular $A^\alpha \subseteq D(\delta)$ and $\delta|_{A^\alpha} = 0$.

§2.8.2 <u>G compact, abelian, $\delta|_{A^\alpha}$ generator</u>

Example 2.8.5 showed that the conditions of the heading of this section do not suffice to ensure that an invariant, closed derivation δ of A is a generator, even when $A = C(\Omega)$ is abelian and unital and $G = \mathbf{T}$. The geometric reason is that along the integral curves in Ω defined by δ points may burst into fibres, and conversely, fibres may merge into points. It is, however, easy to see geometrically that this type of phenomenon cannot occur if the set of points $\omega \in \Omega$ of frequency smaller than or equal to $\frac{1}{n}$ is closed for $n = 1, 2, \dots$ (This set is automatically open by Lemma 2.4.18), i.e. if the ideals $A^\alpha(n)A^\alpha(n)^*$ in A^α has an identity (which is a central projection in A^α) for each $n \in \mathbf{Z}$. More generally, the following theorem is valid.

<u>Theorem 2.8.8</u> (Bratteli - Jørgensen)

Let G be a compact, abelian group and α an action of G on a C^*-algebra A . Assume that each of the ideals $A^\alpha(\gamma)A^\alpha(\gamma)^*$ in A^α

has an approximate identity consisting of projections. Let δ be a closed derivation of A such that

$$[\delta, \alpha] = 0 \qquad \text{(See Definition 2.5.9)}$$

and

$$\delta|_{A^\alpha} \text{ is a generator} \qquad \text{(See 2.5.25)}$$

It follows that δ is a generator.

Remark 2.8.9

In the case that A^α is an AF-algebra, all ideals in A^α have an approximate identity consisting of projections, see [Bra 1]. Thus Theorem 2.8.8 applies to gauge actions of quantum lattice systems, see [BJ 2] for details.

In the special case that all the ideals $A^\alpha(\gamma)A^\alpha(\gamma)^*$ has a common identity, this theorem was proved in [KR 2], and by elaborating the techniques from this paper the present theorem was proved in [BJ 2].

Proof

If A does not have an identity, adjoin one and extend δ and α in the obvious manner. The hypotheses in Theorem 3.1 are then still fulfilled and we may assume from now that A has an identity.

The strategy of the proof is as follows: We prove that the restriction of δ to each of the spectral subspaces $A^\alpha(\gamma)$ is a generator of a group of isometries. Actually, we restrict δ to subspaces of $A^\alpha(\gamma)$ of the form $A^\alpha(\gamma)p$, where p are suitable smooth projections in an approximate identity for $A^\alpha(\gamma)^*A^\alpha(\gamma)$. We then perturb δ by an inner derivation $\delta_{-h} = -\mathrm{ad}(h)$ implemented by an element $h \in A^\alpha$ to obtain a derivation $\delta^{-h} = \delta + \delta_{-h}$ such that $\delta^{-h}(p) = 0$, and hence δ^{-h} leaves $A^\tau(\gamma)p$ invariant. The derivation δ^{-h} still satisfies the same hypotheses as δ, in particular $\delta^{-h}|A^\alpha$ is a generator. We then construct an explicit isometry between $A^\tau(\gamma)p$ and a closed subspace of A^τ. (Actually this is a simplified account, and we have to work in a tensor product.) Transporting δ^{-h} by this isometry, we get an operator on the closed subspace of A^α which is a bounded perturbation of α^{-h}, and therefore is a generator there. It follows that $\delta^{-h}|A(\gamma)p$ is a generator. Perturbing back with a suitable 1-cocycle in A, we obtain the group generated by δ.

Lemma 2.8.10

If $\gamma \in \hat{G}$, there is a net (p_τ) of projections in $A^\alpha(\gamma)^* A^\alpha(\gamma)$ such that:

1. Each $p \in (p_\tau)$ is a finite sum $p = \sum_k x_k^* x_k$, where $x_k \in A^\alpha(\gamma) \cap D(\delta)$.
2. If $x \in A^\alpha(\gamma)$, then $\lim_\tau xp_\tau = x$, where the limit exists in norm.

Proof

The ideal $A^\tau(\gamma)^* A^\tau(\gamma)$ has an approximate identity (p'_β) consisting of projections. By Lemma 2.5.8, $A(\gamma) \cap D(\delta)$ is dense in $A^\alpha(\gamma)$. It follows from the proof of Lemma 2.5.1 that for any $\varepsilon > 0$ there are a finite number of elements $y_k \in A^\alpha(\gamma) \cap D(\delta)$ such that

$$\| p'_\beta - \sum_k y_k^* y_k \| < \varepsilon .$$

If $y = \sum_k y_k^* y_k$, then $y \in A^\tau \cap D(\delta)$, and as p'_β is a projection, the spectrum of y is contained in a small neighbourhood of the set $\{0,1\}$. Let $f \in C^\infty(\mathbb{R})$ be a function which is 0 in a suitable neighbourhood of 0 and $f(x) = x^{-1/2}$ in a suitable neighbourhood of 1, such that $p_\tau = p_{\beta,\varepsilon} = f(y)yf(y)$ is a projection. Then $f(y) \in D(\delta)$ by Theorem 1.6.2 and hence $p_\tau \in D(\delta)$. Furthermore, $\| p'_\beta - p_{\beta,\varepsilon} \|$ is dominated by a constant, only depending on ε, which vanishes as $\varepsilon \to 0$.

Thus, if the set of (β, ε) is ordered by $(\beta, \varepsilon) < (\beta', \varepsilon')$ if $\beta < \beta'$ and $\varepsilon' \leq \varepsilon$, the net p_τ has the property that $\lim xp_\tau = x$, for all $x \in \overline{A^\alpha(\gamma)^* A^\alpha(\gamma)}$. Hence $\lim_\tau yxp_\tau = yx$, for $x \in A^\alpha(\gamma)$, $y \in A^\alpha(\gamma) A^\alpha(\gamma)^*$. But then it follows from Lemma 2.5.1 that $\lim_\tau xp_\tau = x$, for all $x \in A^\alpha(\gamma)$.

If $x_k = y_k f(y)$, then $x_k \in A^\alpha(\gamma) \cap D(\delta)$ and $p_\tau = \sum_k x_k^* x_k$. This ends the proof of Lemma 2.8.10.

From now we fix a $\gamma \in \hat{G}$ and a projection $p = \sum_{k=1}^{N} x_k^* x_k$, where each $x_k \in A^\alpha(\gamma) \cap D(\delta)$. Put $h = \delta(p)p - p\delta(p)$, $\delta_h(x) = hx - xh$, and $\delta^{-h}(x) = \delta(x) - \delta_h(x)$ for $x \in D(\delta)$.

Then h is a skew-adjoint element in A^α, and it follows that δ^{-h} is a closed *-derivation satisfying the same hypotheses as δ in Theorem 3.1. Furthermore $\delta^{-h}(p) = 0$, see Example 1.6.4.

Let δ' be the restriction of δ^{-h} to $D(\delta') = A^\alpha(\gamma)p \cap D(\delta)$. As $\delta^{-h}(p) = 0$ it follows that δ' is a densely defined closed operator on $A^\alpha(\gamma)p$.

Lemma 2.8.11

The operator δ' is the generator of a strongly continuous one-parameter group S_t^p of bounded operators on $A^\alpha(\gamma)p$.

Proof

Recall that $p = \sum_{k=1}^{N} x_k^* x_k$. Consider the C*-dynamical system $(A_N = A \otimes M_N, G, \bar{\alpha})$, where M_N is the $N \times N$ matrix algebra, and $\bar{\alpha}(g) = \alpha(g) \otimes \iota$, where ι is the trivial action. Let $\bar{\delta} = \delta^{-h} \otimes \iota$ with $D(\delta) = D(\delta) \otimes \iota$. Then $\bar{\alpha}$ and $\bar{\delta}$ satisfy the same properties as α and δ.

Define

$$V = \begin{bmatrix} x_1 & 0 & \dots & 0 \\ x_2 & 0 & \dots & 0 \\ \cdot & & & \\ \cdot & & & \\ \cdot & & & \\ x_N & 0 & \dots & 0 \end{bmatrix} \in A_N^{\bar{\alpha}}(\gamma) \cap D(\bar{\delta}) .$$

Then

$$V^*V = \begin{bmatrix} p & 0 & \dots & 0 \\ 0 & 0 & \dots & 0 \\ \cdot & & & \\ \cdot & & & \\ \cdot & & & \\ 0 & 0 & \dots & 0 \end{bmatrix} ,$$

and hence V is a partial isometry, and VV^* is a projection in $A_N^{\bar{\alpha}} \cap D(\bar{\delta})$. Furthermore

$$A_N^{\bar{\alpha}}(\gamma)V^*V = \begin{bmatrix} A^{\alpha}(\gamma)p & 0 \ldots 0 \\ A^{\alpha}(\gamma)p & 0 \ldots 0 \\ \cdot & \cdot \quad \cdot \\ \cdot & \cdot \quad \cdot \\ \cdot & \cdot \quad \cdot \\ A^{\alpha}(\gamma)p & 0 \ldots 0 \end{bmatrix}$$

Thus Lemma 2.8.11 follows once we can show that the restriction $\bar{\delta}'$, of $\bar{\delta}$ to $A_N^{\alpha}(\gamma)V^*V$ is a generator.

To this end, we define an isometric isomorphism ϕ from the Banach space $A_N^{\bar{\alpha}}(\gamma)V^*V$ onto the Banach space $A_N^{\bar{\alpha}}VV^*$ by $\phi(A) = AV^*$, for $A \in A_N^{\bar{\tau}}(\gamma)V^*V$. Then $\phi^{-1}(B) = BV$ for $B \in A_N^{\bar{\tau}}VV^*$, and hence

$$\phi\bar{\delta}'\phi^{-1}(B) = \bar{\delta}'(BV)V^* = \bar{\delta}(BV)V^* .$$

But as $B = BVV^*$, we have by the derivation property $\bar{\delta}(B) = \bar{\delta}(BV)V^* + BV\bar{\delta}(V^*)$. Thus $\phi\bar{\delta}'\phi^{-1}(B) = \bar{\delta}(B) - BV\bar{\delta}(V^*)$, for all $B \in A_N^{\bar{\tau}}VV^*$. Define an operator ρ on $A_N^{\bar{\alpha}}$, by $\rho(B) = \bar{\delta}(B) - BV\bar{\delta}(V^*)$, for $B \in D(\bar{\delta}) \cap A_N^{\bar{\tau}}$. Since $\bar{\delta}$ is the generator of a one-parameter group of automorphisms on $A_N^{\bar{\tau}}$, and the operation of right multiplication by $-V\bar{\delta}(V^*)$ is a bounded operator on $A_N^{\bar{\tau}}$, it follows that ρ is the generator of a strongly continuous one-parameter group, [BR 1], Theorem 3.1.33. The restriction $\phi\bar{\delta}'\phi^{-1}$ of ρ to $A_N^{\bar{\tau}}VV^*$ is then a generator, and as ϕ is an isometry, $\bar{\delta}'$ is a generator. This completes the proof of Lemma 2.8.11.

Next, define a unitary cocycle Γ_t^h in A^{α} by

$$\Gamma_t^h = 1\!\!1 + \sum_{n\geq 1} \int_0^t dt_1 \int_0^{t_1} dt_2 \ldots \int_0^{t_{n-1}} dt_n e^{t_n\delta^{-h}}(h) \ldots e^{t_1\delta^{-h}}(h) ,$$

where $e^{t\delta^{-h}}$ here denotes the one-parameter group of *-automorphisms generated by $\delta^{-h} = \delta - \delta_h$ on A^{τ}. Then $t \to \Gamma_t^h$ is a continuous map into the unitaries in A^{α} satisfying the cocycle relation

$$\Gamma_{t+s}^h = \Gamma_t^h e^{t\delta^{-h}}(\Gamma_s^h) ,$$

and the differential equation

$$\frac{d\Gamma_t^h}{dt} = \Gamma_t^h e^{t\delta^{-h}}(h) ,$$

see Section 2.5.4.

Define a strongly continuous one-parameter family T_t^p of maps from $A^\alpha(\gamma)p$ into $A^\alpha(\gamma)$ by

$$T_t^p(x) = \Gamma_t^h S_t^p(x) \Gamma_t^{h*} .$$

Lemma 2.8.12

If $x \in A^\tau(\gamma)p \cap D(\delta)$, then $T_t^p(x) \in D(\delta)$ for all t, $t \to T_t^p(x)$ is differentiable and

$$\frac{dT_t^p(x)}{dt} = \delta(T_t^p(x))$$

for all $t \in \mathbb{R}$.

Proof

As $x \in D(\delta)$ we have $x \in D(\delta^{-h})$ and hence $x \in D(\delta')$ and

$$\frac{d}{dt} S_t^p(x) = \delta'(S_t^p(x)) = (\delta - \delta_h)(S_t^p(x)) .$$

It follows from Lemma 2.5.18 that $\Gamma_t \in D(\delta^{-h}) = D(\delta)$, and $(d/dt)\Gamma_t^h$ $= \delta^{-h}(\Gamma_t^h) + h\Gamma_t^h = \delta(\Gamma_t^h) + \Gamma_t^h h$. Thus $T_t^p(x) \in D(\delta)$, $t \mapsto T_t^p(x)$ is differentiable and

$$\begin{aligned}
\frac{dT_t^p(x)}{dt} &= \frac{d\Gamma_t^h}{dt} S_t^p(x)(\Gamma_t^h)^* + \Gamma_t^h \frac{dS_t^p(x)}{dt} (\Gamma_t^h)^* + \Gamma_t^h S_t^p(x)\left(\frac{d\Gamma_t^h}{dt}\right)^* \\
&= (\delta(\Gamma_t^h) + \Gamma_t^h h)S_t^p(x)(\Gamma_t^h)^* + \Gamma_t^h(\delta(S_t^p(x)) - h\, S_t^p(x) + S_t^p(x)h)(\Gamma_t^h)^* \\
&\quad + \Gamma_t^h S_t^p(x)(\delta((\Gamma_t^h)^*) - h(\Gamma_t^h)^*) \\
&= \delta(\Gamma_t^h)S_t^p(x)(\Gamma_t^h)^* + \Gamma_t^h\delta(S_t^p(x))(\Gamma_t^h)^* + \Gamma_t^h S_t^p(x)\delta((\Gamma_t^h)^*) \\
&= \delta(T_t^p(x))
\end{aligned}$$

This proves Lemma 2.8.12.

Lemma 2.8.13

If $x \in A^{\alpha}(\gamma)p$, then $\|T_t^p(x)\| = \|x\|$ for all $t \in \mathbb{R}$.

Proof

Let $e^{t\delta}$ again denote the automorphism group generated by the restriction of δ to A^{α} . We first argue that

$$\|T_t^p(x)^* T_t^p(x)\| = \|e^{t\delta}(x^*x)\|$$

for all $t \in \mathbb{R}$. Assume first $x \in A^{\alpha}(\gamma)p \cap D(\delta)$. It follows from Lemma 2.8.12 and the derivation property that the following calculation is valid.

$$T_t^p(x)^* T_t^p(x) - e^{t\delta}(x^*x) = \int_0^t ds \frac{d}{ds}\{e^{(t-s)\delta}(T_s^p(x)^* T_s^p(x))\}$$

$$= \int_0^t ds\; e^{(t-s)\delta}(-\delta(T_s^p(x)^* T_s^p(x)) + \delta(T_s^p(x)^*)T_s^p(x) + T_s^p(x)\delta(T_s^p(x)))$$

$$= \int_0^t ds\; e^{(t-s)\delta}(0) = 0$$

Thus

$$\|T_t^p(x)\|^2 = \|T_t^p(x)^* T_t^p(x)\| = \|e^{t\delta}(x^*x)\| = \|x^*x\| = \|x\|^2 .$$

As $p \in D(\delta)$, $A^{\alpha}(\gamma)p \cap D(\delta)$ is dense in $A^{\alpha}(\gamma)$, and as each T_t^p is bounded, Lemma 2.8.13 follows.

We need the following general lemma.

Lemma 2.8.14 ([BJ 1], Lemma 4.3)

Let G be a compact, abelian group and α on action of G on a C^*-algebra A. Let δ be a closed derivation of A commuting with α and assume that $\delta_0 = \delta\big|_{A^{\alpha}}$ is conservative (Definition 1.6.15). It follows that each of the operators $\pm\delta$ are dissipative on $A^{\alpha}(\gamma)$ for all $\gamma \in \hat{G}$.

Proof

Let $x \in A^{\alpha}(\gamma)$ for some $\gamma \in \hat{G}$. If $x \in D(\delta)$, it follows that

$x^*x \in A^\alpha \cap D(\delta) = D(\delta_0)$, and since δ_0 is conservative, there is a state ω on A^α satisfying $\omega(x^*x) = \|x\|^2$ and $\omega(\delta(x^*x)) = 0$. We extend ω to an α-invariant state on A by the requirement $\omega = \omega \circ P_0$, where $P_0 = \int_G dg\, \alpha_g$.

Define $\phi(y) = \omega(x^*y)$ for $y \in A^\alpha(\gamma)$. Then ϕ is a tangent functional to x in $A^\alpha(\gamma)$, see proof of Theorem 1.4.9. Now

$$\phi(\delta(x)) = \omega(x^*\,\delta(x)) = \omega(\delta(x^*x)) - \omega(\delta(x^*)x) = -\omega(\delta(x^*)x)$$

$$= -\overline{\omega(x^*\,\delta(x))} = -\overline{\phi(\delta(x))},$$

which means that $\phi(\delta(x))$ is purely imaginary: $\mathrm{Re}\ \phi(\delta(x)) = 0$. Thus the two operators $\pm\delta|_{A^\alpha(\gamma)\,\cap\,D(\delta)}$ are dissipative.

We now continue the proof of Theorem 2.8.8.

Lemma 2.8.15

The restriction δ_γ of δ to $A^\alpha(\gamma)$ is the generator of a strongly continuous group of isometries on $A^\alpha(\gamma)$.

Proof

The operators $\pm\delta_\gamma$ are dissipative by Lemma 2.8.14, and by Theorem 1.5.2 it suffices to prove that $(\lambda 1 - \delta)(D(\delta_\gamma))$ is dense in $A^\alpha(\gamma)$ for a positive and a negative real λ. We show this for $\lambda > 0$, the argument for $\lambda < 0$ is analogous. Assume that $x \in A^\alpha(\gamma)p \cap D(\delta)$ where p is as in Lemma 2.8.10. We will show that $x \in \mathrm{range}\ (\lambda 1 - \delta_\gamma)$. Define

$$y = \int_0^\infty dt\ e^{-\lambda t}\ T_t^p(x)$$

where T^p is defined prior to Lemma 2.8.12. The integral converges because of Lemma 2.8.13, and Lemma 2.8.12 implies that

$$\begin{aligned}
\delta(y) &= \int_0^\infty dt\ e^{-\lambda t}\ \delta(T_t^p(x)) \\
&= \int_0^\infty dt\ e^{-\lambda t}\ \frac{d}{dt}(T_t^p(x)) \\
&= \int_0^\infty dt\ \frac{d}{dt}(e^{-\lambda t}\ T_t^p(x)) + \lambda\int_0^\infty dt\ e^{-\lambda t}\ T_t^p(x) \\
&= -x + \lambda y
\end{aligned}$$

Thus $x = (\lambda 1 - \delta)(y)$ and $x \in$ range $(\lambda 1 - \delta_\gamma)$. It now follows from Lemma 2.8.10 that range $(\lambda 1 - \delta_\gamma)$ is dense in $A^\alpha(\gamma)$. This ends the proof of Lemma 2.8.15.

Theorem 2.8.8. is now an immediate consequence of Lemma 2.8.15 and Proposition 2.5.10.

§2.8.3 G compact, A type I, $\delta|_{A^\alpha} = 0$.

It seems plausible that Theorem 2.8.1 also holds when G is not necessarily abelian, but still compact because of Examples 2.8.6 and 2.8.7. One result in this direction is the following

Theorem 2.8.16 (Goodman-Wassermann)

Let G be a compact second countable group and α an action of G on a separable type I C^*-algebra A. Let δ be a closed derivation of A such that

$$[\delta,\alpha] = 0, \quad \text{(See Definition 2.5.9)},$$

the fixed point algebra A^α is contained in $D(\delta)$, and

$$\delta|_{A^\alpha} = 0.$$

It follows that δ is a generator. (See Remark 4.4)

Proof

This was proved in [GW 1] by a non-trivial elaboration of the techniques used in the proof of Theorem 2.4.32, together with the version of the theorem where $A = LC(H)$. This version is Theorem 4.1 in [GJ 1], which is proved as follows: Since any automorphism of $LC(H)$ is implemented by a unitary operator on H which is unique up to a phase factor in $\mathbb{T}$, there is a compact group H, which is an extension of G by $\mathbb{T}$ (i.e. $G = H/\mathbb{T}$) and a unitary representation U of H on H such that $\alpha_g = \mathrm{Ad}(U_h)$ for each $g \in G$, where $h \in H$ is any element which maps into g by the quotient map modulo $\mathbb{T}$. Replacing G by H, we may thus assume that α is unitary implemented by U. But as G is compact, there is an increasing sequence e_n of α-invariant finite rank projections such that the strong limit of e_n as $n \to \infty$ is 1. As e_n is invariant, $e_n \in A^\alpha$ and then $\delta(e_n) = 0$. Thus

$$\delta(e_n x e_n) = e_n \delta(x) e_n$$

for all $x \in D(\delta)$, i.e. δ maps $e_n D(\delta) e_n$ into $e_n A e_n$. But as $e_n A e_n$ is finite-dimensional, it follows that $e_n D(\delta) e_n = e_n A e_n$, and $e_n A e_n$ consists of analytic vectors for δ. If ϕ is any state on A^α, then $\omega = \phi \circ P_0$ is a state on A where $P_0 = \int_G dg\alpha_g$, and as $\delta|_{A^\alpha} = 0$ and $[\delta,\alpha] = 0$ it follows that $\omega \circ \delta = 0$. Thus δ is a generator as a consequence of Corollary 1.5.6.

The general case follows by using a decomposition series of A in closed G-invariant ideals such that Prim $(J/I)/G$ is Hausdorff for any successive ideals $I \subset J$ in the series, and then using imprimitivity characterizations of the C*-algebras over the G-orbits in Prim (J/I). See [GW 1] for details.

§2.9 Noncommutative vector-fields

In this section we will study the classification and generator properties of general derivations $\delta \in \mathrm{Der}\,(A_n^\alpha, A_m^\alpha)$, where $n,m = 0,1,\ldots,\infty,F$, for various special C*-dynamical systems (A,G,α). We will concentrate on results where δ does not satisfy any auxiliary conditions. The special case that δ is invariant, i.e. $[\delta,\alpha] = 0$, was already discussed in §2.6, and some cases where δ is "almost" invariant will be discussed in §2.10. The much more complete theory when A is abelian was discussed in §2.4.

It is one minor exception to the rule that $\delta \in \mathrm{Der}\,(A_n^\alpha, A_m^\alpha)$ should not satisfy additional restrictions in this chapter, occasionally when G is locally compact abelian and $n = F$ we will assume that the restriction of δ to the spectral subspace $A^\alpha(\Omega)$ is bounded for some compact neighbourhood Ω of 0 in $\hat{G}$. This condition implies that $\delta|_{A^\alpha(K)}$ is bounded for all compact $K \subseteq \hat{G}$ by Theorem 2.3.10 and the condition is automatically satisfied when δ is closable. Also, the condition holds if $A_\infty \subseteq D(\delta)$. This is because the generators of the one-parameter semigroups of $\alpha(G)$ are bounded on each $A^\alpha(K)$ where K is compact, Proposition 2.2.14, and hence the Frechét norms defining the topology on A_∞ are equivalent to the operator norm in restriction to each $A^\alpha(K)$. It follows from Theorem 2.3.1 that the restriction of δ to each $A^\alpha(K)$ is bounded, [Kis 2]. Note that when G is compact, $\delta|_{A^\alpha(K)}$ is bounded by Theorem 2.3.8.

We have chosen to organize the multitude of results on noncommutative vector fields in the following way: Some few, but striking, results may be obtained by rather direct computational methods, and these has been collected in §2.9.1. However, the deeper results depends on the existence of G-covariant representations of A with particular properties. This approach has already been initiated in §2.7, and will be completed in §2.9.2. Finally the existence of the relevant representations of A can be deduced from properties of the C^*-dynamical system (A,G,α), and in particular various notions of freeness and ergodicity of the action α plays a prominent role. This is the subject matter of §2.9.3, which therefore contains the main final results of these lecture notes.

§2.9.1 Algebraic theory

§2.9.1.1 Ergodic actions of compact abelian groups

For simplicity of exposition, we first consider the case that A is simple in Theorem 2.9.1, and next the general case in Theorem 2.9.3.

Theorem 2.9.1 (Bratteli-Elliott-Jørgensen)

Let α be an action of a compact abelian group G on a simple C^*-algebra A, and assume that α is ergodic in the sense that $A^\alpha = \mathbb{C}\mathbf{1}$. It follows that all $\delta \in \mathrm{Der}(A_F^\alpha, A)$ has a unique decomposition

$$(2.9.1) \qquad \delta = \delta_0 + \tilde{\delta}$$

where δ_0 is the generator of a one-parameter subgroup of α_G, and $\tilde{\delta}$ is approximately inner.

If $\hat{G}$ is finitely generated (i.e. G is Lie, i.e. $G \cong \mathbf{T}^d \times F$ where $\mathbf{T}$ is the circle group, $d = 0,1,\ldots$, and F is a finite abelian group) and $\delta \in \mathrm{Der}(A_F^\alpha, A_F^\alpha)$, then $\tilde{\delta}$ is inner.

Remark 2.9.2 [BEJ 1], [Con 3], [Con 4, Proposition 49]

It is instructive to consider Theorem 2.9.1 and its proof in the special case $G = \mathbf{T}^2$. Because of the condition $A^\alpha = \mathbb{C}\mathbf{1}$, all the spectral spaces $A^\alpha(\gamma)$, $\gamma \in \hat{\mathbf{T}}^2 = \mathbf{Z}^2$ are one-dimensional and are generated linearly by a unitary operator $U(\gamma)$. Let $U = U((1,0))$ and $V = U((0,1))$. Then A is generated as a C^*-algebra by U and V, and since both

UV and VU are contained in $A^{\alpha}((1,1))$, there is a phase factor $\rho = e^{2\pi i\theta}$, $\theta \in [0,1\rangle$, such that

$$(2.9.2) \qquad VU = \rho UV$$

and this relation, together with $VV^* = V^*V = 1 = UU^* = U^*U$ and $\mathrm{Spec}(U) = \mathrm{Spec}(V) = \mathbf{T}$ determines A uniquely. Any element x in A has a Fourier decomposition

$$(2.9.3) \qquad x = \sum_{nm} c_{nm} U^n V^m$$

where c_{nm} are scalars determined by

$$(2.9.4) \qquad c_{nm} 1 = \frac{1}{(2\pi)^2} \int_0^{2\pi} \int_0^{2\pi} dt_1 dt_2 \, e^{-i(nt_1+mt_2)} \alpha_{(e^{it_1}, e^{it_2})}(x)(V^*)^m(U^*)^n$$

We have that $x \in A_F$ if and only if the sum in 2.9.3 is finite, and $x \in A_\infty$ if and only if the double sequence c_{nm} is rapidly decreasing in the sense that

$$(n,m) \to |c_{nm}|(1 + |n|^k + |m|^k)$$

is bounded for all $k > 0$. In both cases the series 2.9.3 converges in norm.

Any derivation $\delta : A_F^{\alpha} \to A$ is uniquely determined by its action on U and V:

$$(2.9.5) \qquad \begin{aligned} \delta(U) &= \sum_{nm} c^U_{nm} \, U^{n+1} \, V^m \\ \delta(V) &= \sum_{nm} c^V_{nm} \, U^n \, V^{m+1} \end{aligned}$$

Applying δ on both sides of 2.9.2, using the derivation property we obtain the consistency relations

$$(2.9.6) \qquad c^V_{nm}(\rho^m - 1) + c^U_{nm}(\rho^n - 1) = 0$$

If δ is implemented by an operator

$$(2.9.7) \qquad H = \sum_{nm} d_{nm} \, U^n \, V^m,$$

we have the following connection between the c's and the d's:

(2.9.8) $\quad c^U_{nm} = d_{nm}(\rho^m - 1), \quad c^V_{nm} = d_{nm}(1-\rho^n)$

Now, the algebra A is simple if and only if the argument θ is irrational, i.e. if $\rho^n \neq 1$ for all $n \neq 0$. In this case we see that if c^V_{nm}, c^U_{nm} are any pair of double sequences satisfying the consistensy relations 2.9.6, together with the auxilliary conditions

(2.9.9) $\quad c^U_{00} = c^V_{00} = 0,$

then 2.9.8, as a system of equations for d_{nm}, has a solution which is unique up to the value of d_{00}. Now the generators δ_1, δ_2 of the two one-parameter subgroups $t \to \alpha_{(e^{it},0)}$, $t \to \alpha_{(0,e^{it})}$ of $\alpha(\mathbf{T}^2)$ are characterized by

(2.9.10)
$$\delta_1 : \quad c^U_{00} = \sqrt{-1}\ , \text{ all other c-coeffisients are zero.}$$
$$\delta_2 : \quad c^V_{00} = \sqrt{-1}\ , \text{ all other c-coeffisients are zero.}$$

Hence it is clear, at least formally, that any derivation δ has a unique decomposition

(2.9.11) $\quad \delta = \lambda_1\delta_1 + \lambda_2\delta_2 + \tilde{\delta}$

where $\lambda_1 = -ic^U_{00}$, $\lambda_2 = -ic^V_{00}$ and $\tilde{\delta}$ is approximately inner.

Now, assume that $\delta(A^\alpha_F) \subseteq A_\infty$ and that the irrational number θ is badly approximable in the sense that

(2.9.12) $\quad |\theta - \frac{m}{n}| \geq C\, n^{-(2+\varepsilon)}$

for all integers m and $n \neq 0$, and some constants $C, \varepsilon > 0$. By Roth's theorem, [Sch 1], page 116, any irrational algebraic number has this property, and by a theorem of Khintchine, [Sch 1], page 60, this holds for all θ in a subset of $[0,1]$ of Lebesque measure one.

From 2.9.12 it follows that

(2.9.13) $\quad |\rho^n - 1| \geq C'\, n^{-(1+\varepsilon)}$

for some other constant C' and all $n \neq 0$. Thus $(\rho^n - 1)^{-1}$ grows polynomially in n, and as the double sequences c^U_{nm}, c^V_{nm} are rapidly decreasing when $\delta(U), \delta(V) \in A_\infty$, it follows that the solution δ_{nm} of

2.9.8 is rapidly decreasing, and thus $H \in A_\infty$. This, if θ is a badly approximably irrational number, then any approximately inner derivation $\delta \in \mathrm{Der}\,(A_F^\alpha, A_\infty)$ is actually inner, and implemented by an $H \in A_\infty$.

If on the other extreme θ is a Liouville number, there exists approximately inner, non-inner derivations in $\mathrm{Der}\,(A_F^\alpha, A_\infty)$. Another fact which follows from the rapid decrease of c_{nm}^U, c_{nm}^V is that any derivation $\delta \in \mathrm{Der}\,(A_F^\alpha, A_\infty)$ has an extension to a derivation $\delta \in \mathrm{Der}\,(A_\infty, A_\infty)$. The same is true for $G = \mathbb{T}^d$, [BEJ 1], Theorem 4.1.

Proof of Theorem 2.9.1

We follow the proofs of Theorems 1.1 and 2.1 in [BEJ 1].

The proof of the general case goes more or less as the proof of the case $G = \mathbb{T}^2$ which was discussed in Remark 2.9.2. Using the relations 2.2.47,

$$A^\alpha(\gamma)A^\alpha(\gamma)^* \subseteq A^\alpha(0) = \mathbb{C}\,\mathbb{1}\,,$$

$$A^\alpha(\gamma)^*A^\alpha(\gamma) \subseteq A^\alpha(0) = \mathbb{C}\,\mathbb{1}\,,$$

and hence if α is faithful each spectral space $A^\alpha(\gamma)$ is one-dimensional and is spanned by a unitary operator $U(\gamma)$ for all $\gamma \in \hat{G}$. Since both $U(\gamma_1)U(\gamma_2)$ and $U(\gamma_2)U(\gamma_1)$ are eigenunitaries for α corresponding to $\gamma_1+\gamma_2$, there exists a phase factor $\rho(\gamma_1,\gamma_2) \in \mathbb{T} \subseteq \mathbb{C}$ such that

$$(2.9.14)\quad U(\gamma_1)U(\gamma_2) = \rho(\gamma_1,\gamma_2)U(\gamma_2)U(\gamma_1)$$

Clearly $\rho(\gamma_1,\gamma_2)$ is independent of the choice of phase for the U's, and ρ is an antisymmetric bicharacter in γ_1,γ_2. This bicharacter determines the C^*-dynamical system (A,α) uniquely, and conversely any antisymmetric bicharacter $\rho : \hat{G}\times\hat{G} \to \mathbb{T}$ defines an ergodic action of G on a C^*-algebra A, [OPT 1]. As the center of A is α-invariant, it is spanned as a closed linear space by the $U(\gamma)$'s in the center, i.e. by the $U(\gamma)$'s such that $\rho(\gamma,\xi) = 1$ for all $\xi \in \hat{G}$. Also A is simple if and only if the center is trivial, i.e. if and only if the bicharacter is nondegenerate in the sense that $\rho(\gamma,\xi) = 1$ for all $\xi \in \hat{G}$ implies that $\gamma = 0$.

Any element $x \in A$ has a Fourier decomposition

$$(2.9.15)\quad x = \sum_{\gamma\in\hat{G}} c(\gamma)U(\gamma)$$

where the scalars $c(\gamma)$ are given by

$$(2.9.16)\quad c(\gamma)1 = \left\{\int_G dg\ \overline{\gamma(g)}\ \alpha_g(x)\right\} U(\gamma)^*$$

If $\delta : A_F^\alpha \to A$ is a derivation, we have in particular

$$(2.9.17)\quad \delta(U(\gamma)) = \sum_\xi c^\gamma(\xi)U(\gamma)U(\xi)$$

Applying δ to both sides of 2.9.14 we obtain

$$(2.9.18)\quad \delta(U(\gamma_1))U(\gamma_2) + U(\gamma_1)\delta(U(\gamma_2)) = \rho(\gamma_1,\gamma_2)(\delta(U(\gamma_2))U(\gamma_1) + U(\gamma_2)\delta(U(\gamma_1)))$$

Using 2.9.17 and then 2.9.14 on 2.9.18 and comparing Fourier coeffisients on both sides, we obtain the compatibility conditions:

$$(2.9.19)\quad c^{\gamma_1}(\xi)(\rho(\xi,\gamma_2)-1) = c^{\gamma_2}(\xi)(\rho(\xi,\gamma_1)-1),$$

compare with 2.9.6. If δ is implemented by an operator

$$(2.9.20)\quad H = \sum_{\xi\in\hat{G}} d(\xi)U(\xi)$$

we have

$$(2.9.21)\quad \begin{aligned}[H,U(\gamma)] &= \sum_\xi d(\xi)[U(\xi),U(\gamma)]\\ &= \sum_\xi d(\xi)(\rho(\xi,\gamma)-1)U(\gamma)U(\xi)\end{aligned}$$

and hence we get the following connection between the coefficients $c^\gamma(\xi)$ and $d(\xi)$:

$$(2.9.22)\quad c^\gamma(\xi) = d(\xi)(\rho(\xi,\gamma)-1).$$

Now, for a general $\delta : A_F^\alpha \to A$, <u>define</u> δ_0 and $\tilde{\delta}$ by

$$(2.9.23)\quad \delta_0(x) = \int_G dg\ \alpha_g(\delta(\alpha_{-g}(x)))$$

and

$$(2.9.24)\quad \tilde{\delta}(x) = \delta(x) - \delta_0(x),$$

i.e. δ_0 is the invariant part of δ, and $\tilde{\delta}$ the remaining parts.

Then

$$(2.9.25)\quad \begin{aligned} \delta_0(U(\gamma)) &= c^{\gamma}(0)U(\gamma), \\ \tilde{\delta}(U(\gamma)) &= \sum_{\xi \neq 0} c^{\gamma}(\xi)U(\gamma)U(\xi) \end{aligned}$$

for all $\gamma \in \hat{G}$, where we used the normalization $U(0) = 1$.

We now analyze δ_0. The relation $A^{\alpha}(\gamma_1)A^{\alpha}(\gamma_2) \subseteq A^{\alpha}(\gamma_1+\gamma_2)$ between the spectral subspaces implies that there is a complex phase factor $\beta(\gamma_1,\gamma_2)$ such that

$$(2.9.26)\quad U(\gamma_1)U(\gamma_2) = \beta(\gamma_1,\gamma_2)U(\gamma_1+\gamma_2)$$

Applying δ_0 to both sides of this relation and using the derivation property we deduce

$$(2.9.27)\quad c^{\gamma_1}(0) + c^{\gamma_2}(0) = c^{\gamma_1+\gamma_2}(0),$$

i.e. $\gamma \to c^{\gamma}(0)$ is an additive character. As $\delta_0(U(\gamma)^*) = \delta_0(U(\gamma))^*$ we deduce that the scalars $c^{\gamma}(0)$ are purely imaginary. Thus, for each $t \in \mathbf{R}$, the map

$$\gamma \to e^{tc^{\gamma}(0)}$$

is a character, i.e. an element in $\hat{G}$. By Pontryagin's duality theorem, [Rud 1], there is a $g(t) \in G$ such that

$$(2.9.28)\quad e^{tc^{\gamma}(0)} = \gamma(g(t))$$

for all $\gamma \in \hat{G}$. But as $t \to e^{itc^{\gamma}(0)}$ is a character on $\mathbf{R}$, it follows that

$$g(t_1+t_2) = g(t_1) + g(t_2)$$

for all $t_1,t_2 \in \mathbf{R}$, and $t \to g(t)$ is continous by Pontryagin's theorem. Thus

$$t \to \alpha_{g(t)}$$

is a strongly continuous one-parameter group of automorphisms of A, and as

$$\alpha_{g(t)}(U(\gamma)) = \gamma(g(t))U(\gamma) = e^{tc^{\gamma}(0)}U(\gamma),$$

it follows that δ_0 is a pregenerator of this one-parameter subgroup of α_G.

We next show that the derivation $\tilde{\delta}$ is approximately inner. As A is simple, the antisymmetric bicharacter ρ is non-degenerate, i.e. if $\xi \in \hat{G}$ and $\xi \neq 0$, there exists a $\gamma \in \hat{G}$ such that $\rho(\xi,\gamma) \neq 1$. Define

$$(2.9.29) \quad d(\xi) = c^{\gamma}(\xi)(\rho(\xi,\gamma)-1)^{-1},$$

i.e. $d(\xi)$ is the solution of the equation 2.9.22. The compatibility conditions 2.9.19 imply that the $d(\xi)$ defined by 2.9.29 is independent of the particular γ used, and then 2.9.29 is fulfilled for all $\xi,\gamma \in \hat{G}$ with $\xi \neq 0$. If we now formally define

$$(2.9.30) \quad H = \sum_{\xi \neq 0} d(\xi)U(\xi),$$

we formally have the relation

$$(2.9.31) \quad \tilde{\delta}(U(\gamma)) = [H,U(\gamma)]$$

for all $\gamma \in \hat{G}$. We will show that $\tilde{\delta}$ is approximately inner by introducing cutoffs in the expansion 2.9.30, and go to the limit.

Assume first that the group $\hat{G}$ is finitely generated, i.e. isomorphic to $\mathbf{Z}^d \times \hat{F}$ where $\hat{F}$ is a finite abelian group, [Rud 1], and then $G \cong \mathbf{T}^d \times F$. For $N = 1,2,\ldots$ define

$$(2.9.32) \quad \Lambda_N = \left\{\xi \times \phi \in \mathbb{Z}^d \times \hat{F} \mid |\xi_i| \leq N,\ 1 \leq i \leq d\right\},$$

and

$$(2.9.33) \quad \tilde{\delta}_{D,N}(U(\gamma)) = \sum_{0 \neq \xi \in \Lambda_N} c^{\gamma}(\xi)U(\gamma)U(\xi) = \int_G dg\, D_N(g)\alpha_g(\tilde{\delta}(\alpha_{-g}U(\gamma)))$$

where

$$(2.9.34) \quad D_N(g) = \sum_{\xi \in \Lambda_N} \xi(g)$$

is the Dirichlet kernel. Define

$$(2.9.35) \quad \tilde{\delta}_{F,N}(U(\gamma)) = N^{-1} \sum_{n=1}^{N} \tilde{\delta}_{D,n}(U(\gamma)) = \int_G dg\, F_N(g)\alpha_g(\tilde{\delta}(\alpha_{-g}U(\gamma)))$$

where F_N is the Fejér kernel. As the function $g \to \alpha_g(\tilde{\delta}(\alpha_{-g}U(\gamma)))$ is continuous, it follows that

$$(2.9.36)\quad \lim_{N\to\infty} \|\tilde{\delta}_{F,N}(U(\gamma)) - \tilde{\delta}(U(\gamma))\| = 0$$

But

$$(2.9.37)\quad \tilde{\delta}_{F,N} = \mathrm{ad}\,(N^{-1}(H_1+H_2+\ \dots\ +H_N))$$

where

$$(2.9.38)\quad H_n = \sum_{\xi\in\Lambda_n} d(\xi)U(\xi)$$

and it follows from 2.9.36 that $\tilde{\delta}$ is approximately inner.

The general case, where $\hat{G}$ is not necessarily finitely generated, can now be treated by writing $\hat{G}$ as the upward directed union of its finitely generated subgroups Γ, see [BEJ 1], page 172-173 for details.

This shows the existence of the decomposition 2.9.1:

$$\delta = \delta_0 + \tilde{\delta}$$

To show uniqueness it suffices to show that if the generator δ_0 of a one-parameter subgroup of α_G is approximately inner, then $\delta_0 = 0$. Such a δ_0 has the form

$$(2.9.39)\quad \delta_0(U(\gamma)) = c^{\gamma}U(\gamma)$$

where $\gamma \to c^{\gamma}$ is an additive character with values in $i\mathbf{R}$. But if δ' is an inner derivation, it follows from 2.9.22 that the $U(\gamma)$-Fourier component of $\delta'(U(\gamma))$ is zero. Since there exists a projection P_{γ} of norm one from A onto $¢\, U(\gamma)$, 2.2.31, it follows that $c^{\gamma} = 0$ for all γ and $\delta_0 = 0$.

Finally, if $\hat{G}$ is finitely generated and $\delta(A_F^{\alpha}) \subseteq A_F^{\alpha}$, let Λ be a finite generating subset of $\hat{G}$, and put $\Lambda' = \{\xi \in \hat{G} \mid c^{\gamma}(\xi) \neq 0$ for some $\gamma \in \Lambda\}$.

As $\delta(U(\gamma)) \in A_F^{\alpha}$ for $\gamma \in \Lambda$, it follows that Λ' is finite, and

$$H = \sum_{\xi\in\Lambda'} d(\xi)U(\xi) \in A_F^{\alpha}.$$

Then $\tilde{\delta}(U(\gamma)) = [H,U(\gamma)]$ for $\gamma \in \Lambda$ and hence $\tilde{\delta}(U(\gamma)) = [H,U(\gamma)]$ for all $\gamma \in \hat{G}$. This ends the proof of Theorem 2.9.1.

We now consider the case that A is not simple, i.e. the anti-symmetric bicharacter ρ is degenerate. A particular example is $\rho = 1$, then $A \cong C(G)$ and $\alpha \cong$ right (or left) translation.

To formulate the theorem, we need a notion of basis for the Lie algebra of the compact abelian group G, i.e. for the $\mathbf{R}$-linear space of one-parameter groups of G which is naturally identified by Pontryagin duality with the $\mathbf{R}$-linear space $\mathrm{Hom}\,(\hat{G}, i\mathbf{R})$ of group homomorphisms from $\hat{G}$ to $i\mathbf{R}$. We will say that $\{g_k\}_{k\in\Omega}$ is a dual basis for the Lie algebra of G, or for $\mathrm{Hom}\,(\hat{G}, i\mathbf{R})$, if there is a basis $\{\varepsilon_k\}_{k\in\Omega}$ of the $\mathbf{Q}$-linear space $\mathbf{Q}\otimes\hat{G}$ = the divisible hull of $\hat{G}$ such that $g_k(\varepsilon_m) = \delta_{km}$. Then there is an isomorphism

(2.9.40) $$(\beta_k) \in \mathbf{R}^{\Omega} \to \sum_k \beta_k g_k \in \mathrm{Hom}\,(\hat{G}, i\mathbf{R})$$

where the last sum converges in the sense that on each element in $\hat{G}$ it is finite. In particular, if $G = \mathbf{T}^{\Omega}\times H$ where H is a compact abelian torsion group, the canonical one-parameter subgroups of $\mathbf{T}^{\Omega}$ indexed by Ω define a dual basis for the Lie algebra of G.

Theorem 2.9.3 (Bratteli-Elliott-Jørgensen)

Let α be an action of a compact abelian group G on a C^*-algebra A, and assume that α is ergodic in the sense that $A^{\alpha} = \mathbb{C}\mathbb{1}$. It follows that all $\delta \in \mathrm{Der}\,(A_F^{\alpha}, A)$ has a unique decomposition

(2.9.41) $$\delta = \delta_0 + \tilde{\delta}$$

where $\tilde{\delta}$ is an approximately inner derivation, and $\delta_0 : A_F^{\alpha} \to A$ is a $*$-derivation satisfying any of the following four equivalent conditions:

(2.9.42) δ_0 commutes with the action of the subgroup of G acting trivially on the center of A.

(2.9.43) $U(\gamma)^*\delta_0(U(\gamma))$ belongs to the center of A for all $\gamma \in \hat{G}$.

(2.9.44) There is a (unique) map $\gamma \to a(\gamma)$ from $\hat{G}$ to the skewadjoint elements in the center of A such that

$$a(\gamma_1+\gamma_2) = a(\gamma_1) + a(\gamma_2)$$

for all $\gamma_1, \gamma_2 \in \hat{G}$, and

$$\delta_0(U(\gamma)) = a(\gamma)U(\gamma)$$

for all $\gamma \in \hat{G}$.

(2.9.45) If (δ_k) is a dual basis for the Lie algebra of the action of G, then there is a (unique) family (a_k) of self-adjoint elements of the centre of A such that

$$\delta_0 = \sum_k a_k \delta_k$$

where the sum converges in the sense that applied to any element in A_F^α, it is finite.

The additive map $\gamma \to a(\gamma)$ in 2.9.44 and the family (a_k) in 2.9.45 may be arbritary. If $\hat{G}$ is finitely generated, i.e. G is Lie, the basis (δ_k) is finite, and if in addition $\delta \in \mathrm{Der}(A_F^\alpha, A_F^\alpha)$, then $\tilde{\delta}$ is inner.

Proof

This is Theorem 5.1 in [BEJ 1], and the proof is an elaboration of the proof of Theorem 2.9.1, using that the center of A is the closed linear span of the $U(\gamma)$'s such that $\rho(\gamma,\xi) = 1$ for all $\xi \in \hat{G}$. We omit the details.

§2.9.1.2 Analytic elements and condition Γ.

In this section and some of the following we will show under general circumstances that if G is a compact Lie group and $\delta \in \mathrm{Der}(A_F^\alpha, A_F^\alpha)$, then all elements in A_F^α are analytic for δ, see Definition 1.5.3, and so far there does not seem to be any example where this is not true. The basis for this is the following lemma.

Lemma 2.9.4 (Thomsen, Bratteli-Evans)

Let $\mathcal{D}$ be a normed algebra (i.e. $\mathcal{D}$ is an algebra over $\mathbb{C}$ equipped with a norm $\|\ \|$ such that $\|xy\| \leq \|x\|\|y\|$ for all $x,y \in \mathcal{D}$), and assume that $\mathcal{D}$ is generated algebraically by a subalgebra A together with a finite number of elements $s_1,\ldots,s_n$. If $\alpha = (\alpha_1,\ldots,\alpha_m)$ is a multiindex with values $\alpha_i \in \{1,\ldots,n\}$, define $s_\alpha = s_{\alpha_1} s_{\alpha_2} \ldots s_{\alpha_m}$. Let δ be a derivation from $\mathcal{D}$ into $\mathcal{D}$, and assume that for each pair α,β of multiindices there is a bounded linear map $\delta_{\alpha\beta}: A \to A$ such that only a finite number of $\delta_{\alpha\beta}$ are non-zero, and

(2.9.46) $$\delta(x) = \sum_{\alpha\beta} s_\alpha \delta_{\alpha\beta}(x) s_\beta$$

for all $x \in A$.

It follows that all elements $x \in \mathcal{D}$ are analytic for δ, and moreover there exists a $\varepsilon > 0$ such that

$$(2.9.47) \quad \sum_{n=0}^{\infty} \frac{\varepsilon^n}{n!} \|\delta^n(x)\| < +\infty$$

for all $x \in \mathcal{D}$. (We will later say that the elements in $\mathcal{D}$ has a <u>uniform radius of analyticity</u> for δ.)

Furthermore, if $\mathcal{D}_{L,k,i,j}$ is the subset of $y \in \mathcal{D}$ consisting of the linear combinations of at most k monomials of the form

$$s_{\alpha_1} x_1 \, s_{\alpha_2} x_2 \cdots s_{\alpha_m} x_m \, s_{\alpha_{m+1}}$$

where $x_1, \ldots, x_m \in A$, $\alpha_1, \ldots, \alpha_{m+1}$ are multiindices, $m \leq j$, the number of s-factors is less than or equal to i, and the product of the norms of the x-factors is at most L, then the estimate

$$\sum_{n=0}^{\infty} \frac{t^n}{n!} \|\delta^n y\| \leq k\, L\, C_{i,j}(\delta)$$

is valid for all $y \in \mathcal{D}_{L,k,i,j}$ and all $t \in [0,\varepsilon]$, where $c_{i,j}(\delta)$ is independent of k and L.

<u>Remark 2.9.5</u>

The first version of this lemma was proved in [Th 1], while the present version is very similar to that of [BEv 2].

Note that the lemma in particular states that any element in a finitely generated normed algebra is analytic for any derivation on the algebra. In the typical applications of the lemma to actions of a compact group G we will use $\mathcal{D} = A_F^{\alpha}$, $A = A^{\alpha}$, and $s_1, \ldots, s_n$ a finite sequence of elements from the spectral subspaces $A^{\alpha}(\gamma)$, $\gamma \in \hat{G}$.

<u>Proof of Lemma 2.9.4</u>

We follow the proof of [BEv 2, Lemma 2.4]. We may assume that $\|s_i\| = 1$ for $i = 1, \ldots, n$. Any element in $\mathcal{D}$ is a linear combination of monomials of the form

$$(2.9.48) \quad s_{\alpha_1} x_1 \, s_{\alpha_2} x_2 \cdots s_{\alpha_m} x_m \, s_{\alpha_{m+1}}$$

where $x_1,\dots,x_m \in A$, $\alpha_1,\dots,\alpha_{m+1}$ are multiindices and we use the convention that $s_\alpha = 1$ if $\alpha = \emptyset$. The following symbolic notation for elements in $\mathcal{D}$ is useful. Let

(2.9.49) $$\sum_k (i)_s(j)_x L$$

denote any polynomial in $s_1,\dots,s_n$ and $x \in A$ such that the polynomial is a sum of at most k monomials, and each of these monomials is a product of at most i factors s_k and at most j factors in A such that the product of the norms of the latter factors is at most L. In particular $(1)_s$ denotes one of $s_1,\dots,s_n$ and $(1)_x L$ an element in A of norm at most L. Now, it follows from the assumption on $\delta|_A$ that there exists natural numbers N,M,K and a positive number C such that

(2.9.50) $$\delta((1)_x L) = \sum_N (M)_s(1)_x CL.$$

(Actually this is the only place we use the special condition on $\delta|_A$, so we may replace that condition by the formula above.) Since there are only a finite number of generators $s_1,\dots,s_n$, there likewise exist natural numbers N,M,K and a positive number C such that

(2.9.51) $$\delta((1)_s) = \sum_N (M)_s(K)_x\, C,$$

and we may use the same N,M,K,C in the two cases by taking maxima. But using the derivation property $\delta(xy) = \delta(x)y + x\delta(y)$ on monomials, we then have

$$\begin{aligned}
&\delta\Big(\sum_k (i)_s(j)_x L\Big)\\
&= \sum_k \big(\delta((i)_s)(j)_x L + (i)_s\delta((j)_x L)\big)\\
&= \sum_k \sum_{iN} (i+M-1)_s(j+K)_x\, CL\\
&= \sum_k \sum_{jN} (i+M)_s(j+K-1)_x\, CL\\
&= \sum_{k(i+j)N} (i+M)_s(j+K)_x\, CL
\end{aligned}$$

Thus, iterating, we get

$$\delta^2\Big(\sum_k (i)_s(j)_x L\Big)$$

$$= \sum_{k(i+j)(i+j+M+K)N^2} (i+2M)_s (j+2K)_x C^2 L$$

and generally

$$\delta^m(\sum_k (i)_s (j)_x L)$$

(2.9.52)
$$= \sum_{k(i+j)(i+j+M+K)\dots(i+j+(m-1)(M+K))N^m} (i+mM)_s (j+mK)_x C^m$$

Now, using

$$(i+j)(i+j+M+K)\dots(i+j+(m-1)(M+K))$$

$$= (M+K)^m \frac{i+j}{M+K}\left(\frac{i+j}{M+K} + 1\right)\dots\left(\frac{i+j}{M+K} + (m-1)\right)$$

$$\leq (M+K)^m \frac{(D+m)!}{D!}$$

where D is the smallest integer larger than $\frac{i+j}{M+K}$, we get

$$\delta^m(\sum_k (i)_s (j)_x L)$$

(2.9.53)
$$= \sum_{k \frac{(D+m)!}{D!} ((M+K)N)^m} (i+mM)_s (j+mK)_x C^m L$$

This implies the estimate

(2.9.54)
$$\| \delta^m(\sum_k (i)_s (j)_x L) \| \leq kL \frac{(D+m)!}{D!} ((M+K)NC)^m$$

Thus

$$\sum_m \frac{\varepsilon^m}{m!} \| \delta^m(\sum_k (i)_s (j)_x L) \|$$

converges provided

(2.9.55) $\varepsilon < ((M+K)NC)^{-1}$

and the last constant is independent of $\sum_k (i)_s (j)_x L$.

The most immediate application of Lemma 2.9.4 is to actions satisfying a condition Γ:

Definition 2.9.6 [BEv 2]

Let α be an action of a compact group G on a C^*-algebra A, and let $A_n^\alpha(\gamma)$, $\gamma \in \hat{G}$, $n = 1,2,\ldots$ be the corresponding spectral subspaces, see Definition 2.2.12. The action α is said to satisfy condition Γ if A has an identity $\mathbb{1}$ and there exists a finite subset $F \subseteq \hat{G}$ such that

(2.9.56) If $g \in G$ and $\gamma(g) = 1$ for all $\gamma \in F$, then $g = e$.

If $\gamma \in F$ there exists a natural number $n = n(\gamma)$, and a matrix $S(\gamma) \in A_n^\alpha(\gamma)$ such that

(2.9.57) $$S(\gamma)^* S(\gamma) = \mathbf{1}_d$$

where $\mathbf{1}_d$ is the identity in the $d(\gamma) \times d(\gamma)$ matrix algebra over A. (See Remark 4.2)

Note that 2.9.56 implies that G is a Lie group, because G then is isomorphic to a closed subgroup of $U(n)$, where $n = \sum_{\gamma \in F} d(\gamma)$.

Several examples where condition Γ is fullfilled is discussed in the introduction to [BEv 2], in particular when G is abelian the condition is equivalent to the conditions that A is unital, G is Lie and $A^\alpha(\gamma)^* A^\alpha(\gamma) = A^\alpha$ for all $\gamma \in \hat{G}$.

Theorem 2.9.7 (Bratteli-Evans)

Let A be a C^*-algebra with identity, G a compact Lie group, α an action of G on A satisfying the condition Γ.
If $\delta \in \mathrm{Der}\,(A_F^\tau, A_F^\tau)$, then δ is a pre-generator.

This is Theorem 2.1 in [BEv 2], and we follow the proof there. The proof goes via three lemmas.

By definition, there exists a finite subset F of $\hat{G}$ separating points in G such that for each $\gamma \in F$ there is a $S(\gamma) = [s_{ij}^\gamma] \in A_n^\tau(\gamma)$

with $S(\gamma)^*S(\gamma) = \mathbf{1}_n$. Now let $s_1, s_2, \ldots, s_m$, $m = \sum_{\gamma \in F} n(\gamma)d(\gamma)$ be an enumeration of all the matrix elements s^{γ}_{ij}, $i = 1, \ldots, n(\gamma)$; $j = 1, \ldots, d(\gamma)$; $\gamma \in F$. If $\alpha = (i_1, \ldots, i_k)$ is a multiindex, where $i_j \in \{1,2,\ldots,m\}$ for $j = 1, \ldots, k$, define

$$(2.9.58) \quad s_\alpha = s_{i_1} s_{i_2} \ldots s_{i_k}, \text{ and } |\alpha| = k.$$

The first lemma is a characterization of A_F.

Lemma 2.9.8

If $x \in A$, *then* x *is in* A_F *if and only if* x *has the form*

$$(2.9.59) \quad x = \sum_{\alpha\beta} s^*_\alpha x_{\alpha\beta} s_\beta$$

where the sum is over a finite number of multiindices α, β *and* $x_{\alpha\beta} \in A^\tau$ *for all* α, β.

Furthermore, if $\gamma \in \hat{G}$ *there exist nonnegative integers* $k = k(\gamma)$, $\ell = \ell(\gamma)$ *and bounded linear maps* $\sigma^\gamma_{\alpha\beta}: A^\tau(\gamma) \to A^\tau$ *for* $|\alpha| = k$, $|\beta| = \ell$ *such that*

$$(2.9.60) \quad x = \sum_{\substack{|\alpha|=k \\ |\beta|=\ell}} s^*_\alpha \sigma^\gamma_{\alpha\beta}(x) s_\beta$$

for all $x \in A^\tau(\gamma)$. (*This decomposition is not unique.*)

Proof

As $A_F = \oplus_{\gamma \in \hat{G}} A^\tau(\gamma)$, the first statement of the lemma follows from the last. We will construct $\sigma^\gamma_{\alpha\beta}$ by composing and adding a finite number of bounded linear maps.

If $x \in A^\tau(\gamma)$, define (see Definition 2.2.11)

$$x_{ij} = P^\gamma_{ij}(x) = d(\gamma) \int \overline{\gamma_{ji}(g)}\, \tau_g(x)\, dg$$

Then $\| P^\gamma_{ij} \| \leq d(\gamma)$, so the map

$$x \to [x_{ij}]$$

is bounded from $A^{\tau}(\gamma)$ to $A^{\tau}(\gamma) \otimes M_d$. The formula 2.2.39 shows that $[x_{ij}] \in A_d^{\tau}(\gamma)$, and

$$x = \sum_{i=1}^{d} x_{ii}.$$

The representation $\pi = \bigotimes_{\gamma \in F} \gamma$ of G is faithful, and it follows from the Stone-Weierstrass theorem that any coordinate function γ_{ij} of any $\gamma \in \hat{G}$ can be approximated by a linear combination ϕ of the coordinate functions of representations of the form $\bar{\pi}_1 \otimes \bar{\pi}_2 \otimes \ldots \otimes \bar{\pi}_k \otimes \pi_1 \otimes \pi_2 \otimes \ldots \otimes \pi_{\ell}$ $= \bar{\pi}^k \otimes \bar{\pi}^{\ell}$. But if γ were not a subrepresentation of $\bar{\pi}^k \otimes \bar{\pi}^{\ell}$ for any k, ℓ, then $\int \gamma_{ij}(g)\overline{\phi(g)}dg = 0$ for any such function ϕ by Weyl's orthogonality relations. This leads to a contradiction, and hence there exists a pair k, ℓ of nonnegative integers such that γ is a subrepresentation of $\bar{\pi}^k \otimes \pi^{\ell}$. This means there exists a unitary matrix $U \in M_{d(\pi)^{k+\ell}}$, where $d(\pi) = \sum_{\gamma \in F} d(\gamma)$, such that

$$U(\bar{\pi}^k(g) \otimes \pi^{\ell}(g))U^* = \left[\begin{array}{c|ccc} \gamma(g) & 0 & \cdots & 0 \\ \hline 0 & & & \\ \vdots & & \xi(g) & \\ 0 & & & \end{array}\right],$$

where ξ is some unitary scalar matrix. Now, let

$$Y = [y_{ij}] = U^* \left[\begin{array}{c|ccc} [x_{ij}] & 0 & \cdots & 0 \\ \hline 0 & & & \\ \vdots & & 0 & \\ 0 & & & \end{array}\right] U.$$

Now, as

$$\tau_g([x_{ij}]) = [x_{ij}][\gamma_{ij}(g)]$$

and U is a scalar matrix, it follows that

$$\tau_g(Y) = Y(\bar{\pi}^k(g) \otimes \pi^{\ell}(g))$$

Now, Y is an operator valued matrix in $A \otimes M_{d(\pi)^k} \otimes M_{d(\pi)^\ell}$ where $M_{d(\pi)^k}$ corresponds to the representation $\bar{\pi}^k$ and $M_{d(\pi)^\ell}$ to π^ℓ. The transposition map tr of $M_{d(\pi)^k}$ induces a linear map $1 \otimes tr \otimes 1$ of $A \otimes M_{d(\pi)^k} \otimes M_{d(\pi)^\ell}$ which is a permutation of the matrix elements. If Z is the matrix obtained from Y by this map, Z has the transformation law

$$\tau_g(Z) = (\pi^k(g)^* \otimes 1)\, Z\, (1 \otimes \pi^\ell(g))$$

as $\pi^k(g)^*$ is the transpose of $\bar{\pi}^k(g)$.

Next, define the $(\sum_{\gamma \in F} n(\gamma)) \times (\sum_{\gamma \in F} d(\gamma)) = n \times d(\pi)$ matrix $S = [s_{ij}]$ by

$$S = \begin{bmatrix} S(\gamma_1) & 0 & \cdots & 0 \\ 0 & S(\gamma_2) & \cdots & 0 \\ \vdots & \vdots & & \vdots \\ 0 & 0 & \cdots & S(\gamma_r) \end{bmatrix}$$

where $\gamma_1, \ldots, \gamma_r$ is an enumeration of the elements in F. Then

$$S^*S = 1_{d(\pi)}$$

and

$$\tau_g(S) = S\pi(g)$$

It follows from the transformation laws of Z and S that

$$T = [t_{ij}] = ((\overset{k}{\otimes} S) \otimes 1)\, Z\, (1 \otimes (\overset{\ell}{\otimes} S^*))$$

is fixed under τ.

The maps

$$x \to [x_{ij}] \to [y_{ij}] \to [z_{ij}] \to [t_{ij}]$$

are all bounded, and as

$$Z = ((\overset{k}{\otimes} S^*) \otimes 1)\, T\, (1 \otimes (\overset{\ell}{\otimes} S)),$$

Y = reshuffle of the matrix elements of Z,

x_{ij} = linear combination of the matrix elements of Y,

$$x = \sum_{i=1}^{d} x_{ii},$$

it follows that x has the form

$$x = \sum_{\substack{|\alpha|=k \\ |\beta|=\ell}} s_\alpha^* \, \sigma_{\alpha\beta}^{\gamma}(x)\, s_\beta$$

where each $\sigma_{\alpha\beta}^{\gamma}$ is a bounded map from $A^\tau(\gamma)$ into A^τ. This ends the proof of Lemma 2.9.8.

Lemma 2.9.9

Let δ be a derivation from A^τ into A_F. Then there exists a natural number M, and for each pair α,β of multiindices with values in $\{1,2,\ldots,m\}$ a bounded map $\delta_{\alpha\beta}: A^\tau \to A^\tau$ such that

$$(2.9.60)\qquad \delta(x) = \sum_{\substack{|\alpha|\leq M \\ |\beta|\leq M}} s_\alpha^* \, \delta_{\alpha\beta}(x)\, s_\beta$$

for all $x \in A^\tau$. (This decomposition is not unique.)

Proof

The derivation δ is bounded by Lemma 2.5.15, and the rest of the proof only depends on this fact and not on the derivation property of δ.

If M is a natural number, it follows from the proof of Lemma 2.9.8 that the subspace

$$(2.9.61)\qquad A_M = \{ \sum_{\substack{|\alpha|\leq M \\ |\beta|\leq M}} s_\alpha^* \, x_{\alpha\beta}\, s_\beta \mid x_{\alpha\beta} \in A^\tau \}$$

of A_F is exactly the linear span of the finite number of spectral subspaces $A^{\tau}(\gamma)$ where γ is contained in some representation of the form $\bar{\pi}^k \otimes \pi^{\ell}$ with $k \leq M$, $\ell \leq M$. It follows that A_M is closed. Thus $M \to \delta^{-1}(A_M)$ is an increasing sequence of closed subspaces of A^{τ} with union A^{τ}. It follows from Baire's category theorem that $A^{\tau} = \delta^{-1}(A_M)$ for some M, i.e. δ maps A^{τ} into A_M.

Let F_{δ} be the finite set of representations $\gamma \in \hat{G}$ which occurs in the decomposition of τ restricted to A_M. It follows from the proof of Lemma 2.2 that we may choose the k, ℓ occurring there such that $k \leq M$, $\ell \leq M$ if $\gamma \in F_{\delta}$, and we put $\sigma^{\gamma}_{\alpha\beta} = 0$ if $|\alpha| \neq k$, $|\beta| \neq \ell$. Then it follows from Lemma 2.9.8 that

$$\delta(x) = \sum_{\gamma \in F_{\delta}} P_{\gamma}(\delta(x))$$

$$(2.9.62) \qquad = \sum_{\gamma \in F_{\delta}} \sum_{\substack{|\alpha| \leq M \\ |\beta| \leq M}} s^*_{\alpha}\, \sigma^{\gamma}_{\alpha\beta}(P_{\gamma}(\delta(x)))s_{\beta}$$

$$= \sum_{\substack{|\alpha| \leq M \\ |\beta| \leq M}} s^*_{\alpha}\, \delta_{\alpha\beta}(x)s_{\beta}$$

where

$$(2.9.63) \quad \delta_{\alpha\beta}(x) = \sum_{\gamma \in F_{\delta}} \sigma^{\gamma}_{\alpha\beta}(P_{\gamma}(\delta(x)))$$

for all $x \in A^{\tau}$. This ends the proof of Lemma 2.9.9.

Proof of Theorem 2.9.7

It follows from Lemma 2.9.4 and Lemma 2.9.9 that all elements in A^{τ}_F are analytic for δ, with uniform radius ε of convergence, and hence we may define

$$(2.9.64) \quad e^{t\delta}(x) = \sum_{n=0}^{\infty} \frac{t^n}{n!}\, \delta^n(x)$$

if $x \in A^{\tau}_F$ and $|t| < \varepsilon$. It was established in the proof of Lemma 2.9.8 that A_M is exactly the linear span of those spectral subspaces $A^{\tau}(\gamma)$ such that γ is a subrepresentation of $\bar{\pi}^r \otimes \pi^{\ell}$ for some $r \leq M$, $\ell \leq M$. Thus it follows from Lemma 2.9.4 that if Λ is a finite subset of $\hat{G}$, then $e^{t\delta}\big|_{A^{\tau}(\Lambda)}$ is bounded when $|t| < \varepsilon$. It follows from the derivation property of δ that each $e^{t\delta}$ is a *-morphism from A^{τ}_F into A, and in particular

(2.9.65) $e^{t\delta}(x^*x) \geq 0$

for all $x \in A_F$. It now follows from Lemma 2.5.4 that $e^{t\delta}$ is bounded on A_F^τ, and extends by closure to a *morphism of A. But as *morphisms are automatically contractive, it follows by differentiation that δ is well-behaved on A_F^τ. The derivation δ is then a pregenerator by Theorem 1.5.4.

§2.9.2 Spatial theory

In this section we will study generator and classification results for derivations $\delta \in \mathrm{Der}\,(A_F^\alpha, A_n^\alpha)$ which depends on the existence of covariant representations of the C^*-dynamical system (A,G,α) with particular properties, Definition 2.7.1. Fundamental for the proofs of most of these results is Theorem 2.3.12, which states that δ has a natural extension to the σ-weak closure of A in the covariant representation.

§2.9.2.1 Weakly inner representations and abelian groups

Theorem 2.9.10 (Batty-Ikunishi-Kishimoto)

Let G be a locally compact abelian group, α an action of G on a C^*-algebra A, and assume that there exists a faithful G-covariant representation π of A such that each α_g is implemented by a unitary operator $u_g \in M = \pi(A)''$ such that u_g is contained in the fixed point algebra for the extension $\hat{\alpha}$ of α to M, i.e.

(2.9.66) $u_g u_h = u_h u_g$ for all $g,h \in G$, see Definition 2.7.1.

Let $\delta \in \mathrm{Der}(A_F^\alpha, A)$, and (if G is non-compact) assume that $\delta|_{A^\alpha(\Omega)}$ is bounded for a compact neighbourhood Ω of 0 in $\hat{G}$.

It follows that δ is a pregenerator. Furthermore, there exists a constant $C > 0$ such that for all $f \in L^1(G)$ with $\hat{f}(0) = 0$ (here $\hat{f}$ is the Fourier transform of f) one has

(2.9.67) $\|\delta_f\| \leq C\|f\|_1$

where

$$(2.9.68)\quad \delta_f(x) = \int_G dg\, f(g)\, \alpha_g\, \delta\alpha_{-g}(x).$$

Thus, if G is compact, δ has the decomposition

$$(2.9.69)\quad \delta = \delta_{inv} + \tilde{\delta}$$

where δ_{inv} commutes with α and $\tilde{\delta}$ is bounded.

Remark 2.9.11

This theorem was first proved in [BaK 1] under the assumptions that $G = \mathbf{R}^d$ and $D(\delta) = A_1$, then in [Kis 2] under the assumption that π is a direct sum of irreducible representations and this final version was proved in [Iku 2] as a corollary of the proof in [Kis 2] and Theorem 2.3.12. We will follow this proof.

One remarkable feature with Theorem 2.9.10 is that nothing is assumed about the range $\delta(A_F^\alpha)$, in contrast to the case where A is abelian, see e.g. Example 2.4.23 and Theorem 2.4.24. We will see later that this is a common feature in several situations where A is essentially nonabelian. Note that the hypothesis on α in Theorem 2.9.10 implies that $\alpha = id$ if A is abelian.

An elaboration of Example 2.6.7, given in [Kis 2], shows that the hypotheses of Theorem 2.9.10 does not imply that A_∞, or not even the analytic elements for the action α, are necessarily contained in $D(\bar{\delta})$ when $G = \mathbf{T}$.

The paper [Iku 2] contains some other extensions and elaborations of Theorem 2.9.10.

Proof of Theorem 2.9.10

We will prove Theorem 2.9.10 via 5 lemmas. Assume that A is represented on a Hilbert space as in the statement of the theorem, and suppress the notation π. Put $M = A''$, and use α also to denote the σ-weakly continous extension of α to M, given by

$$(2.9.70)\quad \alpha_g(x) = u_g\, x\, u_g^*$$

for $x \in M$, $g \in G$. By Theorem 2.3.12, δ has an extension to A_F^α, which we also denote by δ, such that the restriction $\delta|_{M^\alpha(K)}$ is bounded and σ-weakly continuous for each compact $K \subseteq \hat{G}$. Define

$$(2.9.71) \quad C_0 = \inf_\Omega \| \delta|_{A^\alpha(\Omega)} \|$$

where the infimum is over all compact neighbourhoods of 0 in $\hat{G}$.

Lemma 2.9.12

There exists a skew-adjoint $h \in M$ such that $\|h\| \leq C_0$ and $\delta_1 = \delta - \mathrm{ad}(h)$ commutes with α. Furthermore, if there is compact symmetric neighbourhood Ω of 0 in $\hat{G}$ such that $\delta(A^\alpha(K)) \subseteq A^\alpha(K+\Omega)$ for any compact K, then h may be chosen from $M^\alpha(\Omega)$.

Proof

This is Lemma 4 in [Kis 2]. Let G be the von Neumann algebra generated by $\{u_g | g \in G\}$. Then G is an abelian von Neumann subalgebra of M^α by 2.9.70, and hence $G \subseteq D(\delta)$ and

$$(2.9.72) \quad \| \delta|_G \| \leq \| \delta|_{M^\alpha} \| \leq C_0$$

by 2.3.8 and 2.9.71. It follows from Lemma 2.5.16 that there is a h in the σ-weakly closed convex hull of $\{\delta(u)u^* | u \text{ is a unitary in } G\}$ which implements $\delta|_G$. Replacing h by $\frac{1}{2}(h-h^*)$ and using $M^\alpha(K) = \bigcap_V \overline{A^\alpha(K+V)}$, where the closures are in the σ-weak topology and the intersection is over all compact neighbourhoods of 0 in $\hat{G}$, Lemma 2.9.12 follows.

Lemma 2.9.13

The derivation δ_1 is σ-weakly closable, and its closure generates a one-parameter group of automorphisms of M which commute with α.

Proof

This is Lemma 5 in [Kis 2]. Let M_0 be the norm-closure of M_F^α. Then it follows essentially by Theorem 2.6.2 that the closure of $\delta_1|_{M_F^\alpha}$ generates a one-parameter group β of *-automorphisms of M_0, as one can replace the condition that δ_1 is closed in Theorem 2.6.2 by the condition that $\delta_1\big|_{A^\alpha(K)}$ is bounded for each compact $K \subseteq \hat{G}$, with essentially the same proof.

We have to show that each β is continuous in the σ-weak topology. Let x_τ be a bounded net in M_0 such that $x_\tau \to 0$ and $\beta_t(x_\tau) \to y$ in the σ-weak topology. Then for any $f \in L^1(G)$ such that $\text{supp}\,\hat{f}$ is compact we have

$$\int_G dg\, f(g)\alpha_g(\beta_t(x_\tau)) = \beta_t(\int_G dg\, f(g)\alpha_g(x_\tau))$$

The right side converges σ-weakly to zero as $\tau \to \infty$ since $\beta_t = e^{t\delta}$ is σ-weakly continuous on each bounded set of $M^\alpha(K)$. It follows that

$$\int_G dg\, f(g)\alpha_g(y) = 0$$

for all $f \in L^1(G)$ with $\text{supp}\,\hat{f}$ compact, and hence $y = 0$. Thus β_t is σ-weakly continuous on the unit sphere of M_0, and by Kaplansky's density theorem β_t extends to an automorphism of M, also denoted by β_t. Now, it is easily seen by regularization that $t \to \omega\beta_t$ is norm-continuous for $\omega \in M_*$ with compact α_*-spectrum, and as such ω's are norm dense, we conclude that β_* is strongly continuous on M_*, i.e. $t \to \beta_t$ is σ-weakly continuous.

As the generator of the extension of β to M is an extension of δ_1, and δ_1 has a dense set of analytic vectors, namely M_F^α, it follows that δ_1 is σ-weakly closable and its σ-weak closure is the generator of the extended β, [BR 1], Theorem 3.2.51.

Lemma 2.9.14

The derivation δ is conservative.

Proof

This is Lemma 6 in [Kis 2]. Since δ is a bounded perturbation of δ_1 by Lemma 2.9.12, it follows from Lemma 2.9.13 that δ is σ-weakly closable, and its σ-weak closure is the generator of a one-parameter group of *-automorphisms of M, [BR 1], Theorem 3.1.33. It follows that

$$\|(1-\lambda\delta)(x)\| \geq \|x\|$$

for $x \in M_F^\alpha$ and $\lambda \in \mathbf{R}$, and then in particular for $x \in A_F^\alpha$. Thus δ is conservative by Proposition 2.3.4.

Lemma 2.9.15

Let f be a real function in $L^1(G)$ such that $\operatorname{supp} \hat{f}$ is compact, and define a *-derivation δ_f on A_F^α by 2.9.68. Then any element in A_F^α is entire analytic for δ_f and δ_f is a pregenerator on A.

Proof

This is Lemma 8 and Corollary 9 in [Kis 2]. If $\Omega = \operatorname{supp} \hat{f}$, then $\delta_f(A^\alpha(K)) \subseteq A^\alpha(K+\Omega)$ for all compact $K \subseteq \hat{G}$, and it follows from Lemma 2.9.12 that the extension of δ_f to M_F^α has a decomposition

$$\delta_f = \delta_1 + \operatorname{ad}(h)$$

where δ_1 is α-invariant and $h = -h \in M^\alpha(\Omega \cup (-\Omega))$. Then as $\delta_1\big|_{M^\alpha(K)}$ is bounded and $M^\alpha(K)$ is δ_1-invariant, M_F^α consists of entire analytic elements for δ_1, and in particular h is entire analytic for δ_1. It follows from Lemma 2.5.19 that all elements in M_F^α, and in particular in A_F^α, are entire analytic for δ_f. As δ_f is conservative by Lemma 2.9.14, it follows from Theorem 1.5.4 that δ_f is a pre-generator.

Lemma 2.9.16

Let f be a real function in $L^1(G)$ with $\hat{f}(0) = 0$. Then δ_f is bounded and

$$\|\delta_f\| \leq 4C_0\|f\|_1. \tag{2.9.73}$$

Proof

This is Lemma 10 in [Kis 2]. Since f can be approximated in $L^1(G)$ with functions k such that $\hat{k} = 0$ in a neighbourhood of 0 and $\operatorname{supp} \hat{k}$ is compact, [Rud 1], it suffices to prove (2.9.73) under the latter assumptions. By Lemma 2.9.12, the extension of δ_f to M_F^α has a decomposition

$$\delta_f = \delta_1 + \operatorname{ad}(h) \tag{2.9.74}$$

where δ_1 commutes with α and

(2.9.75) $\|h\| \leq C_0\|f\|_1$

Let $k \in L^1(G)$ be a real function such that $\operatorname{supp} \hat{k}$ is compact and disjoint from $\operatorname{supp} \hat{f}$, and $\hat{k} = 1$ in a neighbourhood of 0 in $\hat{G}$, and $\|k\|_1 < 1+\varepsilon$ for a given $\varepsilon > 0$, [Rud 1, Theorem 2.6.1].

As δ_1 is α-invariant, we have

$$\delta_{1,k} = \int_G dg\, k(g)\alpha_g \delta_1 \alpha_{-g} = \delta_1$$

and since the spectrum of δ_f with respect to the action $g \to \operatorname{Ad}(\alpha_g)$ is contained in $\operatorname{supp} \hat{f}$, it follows that

$$\delta_{f,k} = \int_G dg\, k(g)\alpha_g\, \delta_f\, \alpha_{-g} = 0$$

and hence, from 2.9.74

$$\delta_1 = -\operatorname{ad}(h)_k = -\operatorname{ad}(\alpha_k(h))$$

Thus

$$\|\delta_1\| \leq 2\|\alpha_k(h)\| \leq 2\|k\|_1\|h\| \leq 2C_0(1+\varepsilon)\|f\|_1$$

where we used 2.9.75. As ε was arbitrary we get

$$\|\delta_1\| \leq 2C_0\|f\|_1$$

Thus

$$\|\delta\| \leq \|\delta_1\| + \|\operatorname{ad}(h)\| \leq 4C_0\|f\|_1$$

<u>Proof of Theorem 2.9.10</u>

Let f be a real function in $L^1(G)$ such that $\operatorname{supp} \hat{f}$ is compact and $\hat{f}(0) = 1$. Then for any $g \in L^1(G)$ we have

$$\hat{g}(0) - \hat{f}(0)\hat{g}(0) = 0$$

and hence, by Lemma 2.9.16

$$\begin{aligned}\|(\delta-\delta_f)_g\| &= \|\delta_{g-f*g}\| \\ &\leq 4C_0\|g-f*g\|_1 \\ &\leq 4C_0(\|g\|_1 + \|f\|_1\|g\|_1) \\ &= 4C_0(1+\|f\|_1)\|g\|_1\end{aligned}$$

where $*$ denotes convolution product. It follows that $\delta-\delta_f$ is bounded, and

(2.9.76) $\|\delta-\delta_f\| \leq 4C_0(1+\|f\|_1)$

Thus δ is a bounded perturbation by δ_f. As δ_f is a pregenerator by Lemma 2.9.15, it follows that δ is a pregenerator, [BR 1], Theorem 3.1.33.

If G is compact, we put $\delta_{inv} = \delta_f$ with $f = 1$ in 2.9.76 to obtain 2.9.69. This ends the proof of Theorem 2.9.10.

In Theorem 2.9.10 it is not clear whether δ is always a bounded perturbation of an invariant derivation, but as noted in Remark 2.9.11 δ is not necessarily a bounded perturbation of the generator of a one-parameter subgroup of α_G. However:

Corollary 2.9.17 (Batty-Kishimoto)

Adopt the assumptions of Theorem 2.9.10, but assume also that A has a faithful G-covariant factor representation ρ (necessarily disjoint from π if G is not finite) such that $\hat{G}/\Gamma(\bar{\alpha})$ is compact, where $\Gamma(\bar{\alpha})$ is the Connes spectrum of the extension $\bar{\alpha}$ of α to $\rho(A)''$.

It follows that δ has the (unique) decomposition

$$\delta = \delta_0 + \tilde{\delta}$$

on A_F^α, where δ_0 is the generator of a one-parameter subgroup of α_G and $\tilde{\delta}$ is bounded.

Proof

This is a minor modification of Theorem 3.6 in [BaK 1].

Let $\{\Omega_\tau\}$ be a neighbourhood basis of 0 in $\hat{G}$ consisting of precompact neighbourhoods, and for each τ let f_τ be a positive function in $L^1(G)$ such that $\int_G dg\, f_\tau(g) = 1$ and $\operatorname{supp} \hat{f}_\tau \subseteq \Omega_\tau$. Let

$$\delta_\tau = \delta_{f_\tau} = \int_G dg\, f_\tau(g)\, \alpha_g \circ \delta \circ \alpha_{-g}$$

By 2.9.76,

$$\|\delta - \delta_\tau\| \leq 8C_0$$

for all τ.

If $N = \rho(A)''$, it follows from Theorem 2.3.12 that δ, δ_τ and $\delta - \delta_\tau$ extends to derivations $\hat{\delta}$, $\hat{\delta}_\tau$ and $\tilde{\delta}_\tau$ on $N_F^{\bar{\alpha}}$ such that the restrictions to each spectral subspace $N^{\bar{\alpha}}(K)$ are bounded and σ-weakly continuous, and it follows from Kaplanskys density theorem and Lemma 2.5.15 that

$$\|\tilde{\delta}_\tau\| \leq 8C_0 .$$

Let $\tilde{\delta}_1$ be a σ-weak limit point of $\tilde{\delta}_\tau$ as $\tau \to \infty$, and define $\tilde{\delta}_{inv}$ on $N_F^{\bar{\alpha}}$ by

$$\hat{\delta}_{inv} = \hat{\delta} - \tilde{\delta}_1$$

Then

$$\hat{\delta}_{inv}(x) = \lim_\tau \hat{\delta}_\tau(x) = \lim_\tau \int_G dg\, f_\tau(g) \alpha_g\, \delta\, \alpha_{-g}(x)$$

for $x \in N_F^{\bar{\alpha}}$, where the limit is in the σ-weak topology over a subnet. But as $\operatorname{supp} \hat{f}_\tau \to 0$ it follows from [BR 1, Proposition 3.2.40] that $\hat{\delta}_{inv}$ is $\bar{\alpha}$-invariant. One deduces as in Lemma 2.9.13 that $\hat{\delta}_{inv}$ is σ-weakly closable, and its closure is the generator of a one-parameter group β of automorphisms of N commuting with α. Now one uses the condition $\hat{G}/\Gamma(\bar{\alpha})$ exactly as in the proof of Theorem 2.6.6 to deduce that $\hat{\delta}_{inv}$ has the decomposition

$$\hat{\delta}_{inv} = \hat{\delta}_0 + \tilde{\delta}_2$$

where $\hat{\delta}_0$ is the generator of a one-parameter subgroup of $\bar{\alpha}_G$ and $\hat{\delta}_2$ is bounded. But then

$$\hat{\delta} = \hat{\delta}_0 + \tilde{\delta}_2 + \tilde{\delta}_1 = \hat{\delta}_0 + \tilde{\delta}$$

where $\tilde{\delta}$ is bounded. As $\hat{\delta}_0(\rho(A_F^\alpha)) \subseteq \rho(A_F^\alpha)$ it follows that $\tilde{\delta}$ defines a bounded derivation in $\mathrm{Der}\,(A,A)$, and Corollary 2.9.17 follows.

§2.9.2.2 Compact abelian Lie groups and $[A^\alpha(\gamma)H] = H$.

Theorem 2.9.18 (Bratteli-Goodman-Jørgensen-Thomsen)

Let G be a compact abelian Lie group, α an action of G on a C^*-algebra A, and assume that A may be faithfully and covariantly represented on a Hilbert space H such that if

$$(2.9.77)\quad \begin{aligned}\Gamma &= \{\gamma \in \hat{G} \mid A^\alpha(\gamma)A^\alpha(\gamma)^* \text{ is } \sigma\text{-weakly dense in } M^\alpha = (A^\alpha)''\} \\ &= \{\gamma \in \hat{G} \mid [A^\alpha(\gamma)H] = H\}\end{aligned}$$

then $\hat{G}/(\Gamma \cap (-\Gamma))$ is finite. Let $\delta \in \mathrm{Der}\,(A_F^\alpha, A_F^\alpha)$.

It follows that δ is a pregenerator. Furthermore there exists an $\varepsilon > 0$ such that

$$(2.9.78)\quad \sum_{n=0}^{\infty} \frac{\varepsilon^n}{n!} \|\delta^n(x)\| < +\infty$$

for all $x \in A_F^\alpha$.

Remark 2.9.19

This theorem was developed successively in [BGJ 1], [Tho 1] and [BG 1], and the present version is a slight generalization of Theorem 3.4 in [BG 1].

Note that the subset $\Gamma \subseteq \hat{G}$ defined by 2.9.77 is a sub-semigroup as a consequence of 2.2.47, and hence $\Gamma \cup (-\Gamma)$ is a subgroup. If $\Gamma(\alpha)$ denotes the Connes spectrum of the extension of α to $M = A''$, Definition 2.6.3, then the hypothesis of the theorem is in particular fulfilled with $\Gamma = \hat{G}$ if $\Gamma(\alpha) = \hat{G}$, in fact the single condition $\Gamma(\alpha) = \hat{G}$ is equivalent to the two conditions

$$(2.9.79)\quad \Gamma = \hat{G}$$

and

$$(2.9.80)\quad M^\alpha \cap (M^\alpha)' \subseteq M \cap M',$$

see [BEv 1, Remark 4.9].

It is essential for the proof of Theorem 2.9.18 that the discrete group $\hat{G}$ is finitely generated, i.e. G is Lie, and the theorem is not longer true if this condition is removed, see Example 2.9.21.

In the proof of Theorem 2.9.18, and later, we will need the following general lemma.

Lemma 2.9.20 (Bratteli-Goodman-Jørgensen)

Let M be a von Neumann algebra, and α a (σ-weakly continuous) action of a compact group G on M. Let $\delta : M^{\alpha} \to M_F^{\alpha}$ be a derivation.

It follows that there exists a $h \in M_F^{\alpha}$ such that

(2.9.81) $\delta(x) = [h,x]$

for all $x \in M^{\alpha}$.

Proof

This was proved when G is abelian in [BGJ 1], and in the general case in [BG 1].

We first argue that there exists a finite subset $K \subseteq \hat{G}$ such that

(2.9.82) $\delta(M^{\alpha}) \subset M^{\alpha}(K)$.

If this were false, there would exist a sequence γ_i of distinct elements in $\hat{G}$ such that

(2.9.83) $P_{\gamma_i}(\delta(A^{\alpha})) \neq \{0\}$

for all i, where P_{γ} is the canonical projection from M into $M^{\alpha}(\gamma)$, Definition 2.2.1. The operator

(2.9.84) $P = \sum_i P_{\gamma_i}$

makes sense as a linear operator on M_F^{α}, and then $P \circ \delta$ is well defined and is a derivation from M^{α} into the linear span of the spaces $M^{\alpha}(\gamma_i)$, $i = 1,2,\ldots$. By Lemma 2.5.15, $P \circ \delta$ is continuous, and hence

(2.9.85) $M_n = \{x \in M^{\alpha} | P \circ \delta(x) \in \mathrm{span}\{A^{\alpha}(\gamma_i) | 1 \leq i \leq n\}\}$

is an increasing sequence of proper closed subspaces of M^{α}, with union equal tc M^{α}. But this contradicts Baire's category theorem, and this contradiction establishes that

(2.9.86) $K = \{\gamma \in \hat{G} | P_{\gamma} \circ \delta \neq 0\}$

is finite. Thus 2.9.82 follows, with K given by 2.8.86.

Now, fix $\gamma \in \hat{G}$ of dimension δ, and define

(2.9.87) $\pi(x) = x \otimes \mathbb{1}_d$

for $x \in M^{\alpha}$, where $\mathbb{1}_d$ is the identity operator on $\mathbb{C}^d$. Let

$$P^{\alpha}_{ij}(\gamma) = d \int_G dg\, \overline{\gamma_{ji}(g)}\, \alpha_g$$

be the maps of Definition 2.2.11, and define maps Δ_i on M^{α} by

(2.9.88) $\Delta_i(x) = \begin{bmatrix} P^{\alpha}_{i1}(\gamma)(\delta(x)) \\ \vdots \\ P^{\alpha}_{id}(\gamma)(\delta(x)) \end{bmatrix}$

for $i = 1,\dots,d$. These maps satisfy the derivation identity

(2.9.89) $\Delta_i(xy) = \pi(x)\Delta_i(y) + \Delta_i(x)y$

for $x,y \in M^{\alpha}$, and also

(2.9.90) $\Delta_i(x)^*\Delta_i(y) \in M^{\alpha}$

for all $x,y \in M^{\alpha}$, since $\Delta_i(x)^* \in M^{\alpha}_1(\overline{\gamma})$. Therefore Theorem 2.5.11 implies that there is an element

(2.9.91) $h_i = \begin{bmatrix} h_{i1} \\ \vdots \\ h_{id} \end{bmatrix} \in M^{\alpha}_1(\gamma)^T,$

where ^{T} denotes transpose, such that

(2.9.92) $\Delta_i(x) = h_i x - \pi(x)h_i$

for $x \in M^{\alpha}$. In particular

(2.9.93) $P_{\gamma}\delta(x) = \sum_i P_{ii}(\gamma)(\delta(x)) = [\sum_i h_{ii}, x]$

for $x \in M^{\alpha}$. Define $h(\gamma) = \sum_i h_{ii}$. Then it follows from 2.9.82 that δ is implemented by

$$(2.9.94)\quad h = \sum_{\gamma\in K} h(\gamma)$$

This ends the proof of Lemma 2.9.20.

Proof of Theorem 2.9.18

Our aim is to show that there exists an $\varepsilon > 0$ such that

$$(2.9.95)\quad \sum_{n=0}^{\infty} \frac{\varepsilon^n}{n!} \|\delta^n(x)\| < +\infty$$

for all $x \in A_F^\alpha$, and, moreover, for each finite $K \in \hat{G}$ a constant C_K such that

$$(2.9.96)\quad \sum_{n=0}^{\infty} \frac{\varepsilon^n}{n!} \|\delta^n(x)\| \leq C_K\|x\|$$

for all $x \in A^\alpha(K)$.

By Theorem 2.3.8 or 2.3.12, δ extends uniquely to a derivation from M_F^α into M_F^α such that $\delta\big|_{M^\alpha(K)}$ is bounded and σ-weakly continuous for each finite $K \subseteq \hat{G}$. We denote also this extension by δ. By Lemma 2.9.20 there exists an $h \in M_F^\alpha$, which we may take to be skew-adjoint, such that $\delta\big|_{M^\alpha} = \mathrm{ad}(h)$. Put

$$(2.9.97)\quad \tilde{\delta} = \mathrm{ad}(h), \quad \delta_0 = \delta - \tilde{\delta}$$

Then δ_0 and $\tilde{\delta}$ maps M_F^α into M_F^α and

$$(2.9.98)\quad \delta_0\big|_{M^\alpha} = 0$$

Let $E(\gamma)$ be the range projection of the σ-weak closure of the ideal $M^\alpha(\gamma)M^\alpha(\gamma)^*$ in M^α. By Lemma 2.7.5 there exists for each $\gamma \in \hat{G}$ an operator $L(\gamma)$ in the relative commutant of $M^\alpha E(\gamma)$ in $E(\gamma)ME(\gamma)$ such that

$$(2.9.99)\quad \delta_0(x) = L(\gamma)x$$

for all $x \in M^\alpha(\gamma)$.

We showed in the proof of Lemma 2.7.5 that for each $\gamma \in \hat{G}$ there is a maximal partial isometry $u(\gamma)$ in $M^\alpha(\gamma)$, maximality being charac terized by the property

$$(2.9.100)\quad (E(\gamma) - u(\gamma)u(\gamma)^*)M^\alpha(\gamma)(E(-\gamma) - u(\gamma)^*u(\gamma)) = 0$$

and then the central support of $u(\gamma)u(\gamma)^*$ in M^α is $E(\gamma)$ and the central support of $u(\gamma)^*u(\gamma)$ is $E(-\gamma)$. Now, if K is a Hilbert space whose dimension has sufficiently high cardinality, then any projection in $M^\alpha \otimes L(K)$ of the form $p \otimes \mathbf{1}$ is equivalent to a central projection in $M^\alpha \otimes L(K)$. Redefining α as $\alpha \otimes 1$, we thus embed our original dynamical system (M,G,α) into a dynamical system $(M \otimes L(H),G,\alpha \otimes 1)$, and in the new dynamical system we can choose $u(\gamma)$ with $u(\gamma)u(\gamma)^* = E(\gamma)$, $u(\gamma)^*u(\gamma) = E(-\gamma)$, where $E(\gamma)$ now means $E(\gamma) \otimes \mathbf{1}$. Now $\delta_0 \otimes 1$ will be defined as an operator on $(M \otimes L(H))_F^{\alpha \otimes 1}$ since the restriction to the spectral subspace $(M \otimes L(H))^{\alpha \otimes 1}(\gamma) = M^\alpha(\gamma) \otimes L(H)$ is given by left multiplication by the bounded operator $L(\gamma) \otimes \mathbf{1}$. Also $\tilde{\delta} \otimes 1 = \mathrm{ad}(h \otimes \mathbf{1})$ is a bounded derivation, and thus $\delta \otimes 1 = \delta_0 \otimes 1 + \tilde{\delta} \otimes 1$ is defined as a derivation from the algebra of G-finite elements in $M \otimes L(H)$ into itself. It suffices to prove 2.9.95 and 2.9.96 for this new derivation. In short we may assume there exists an

$$(2.9.101)\quad u(\gamma) \in M^\alpha(\gamma)$$

such that

$$(2.9.102)\quad u(\gamma)u(\gamma)^* = E(\gamma), \quad u(\gamma)^*u(\gamma) = E(-\gamma)$$

for each $\gamma \in \hat{G}$.

Now, as G is Lie, $\hat{G}$ is finitely generated, i.e. $\hat{G}$ is the product of a lattice $\mathbf{Z}^m$ and a finite group, [Rud 1], and hence the subgroup $\Gamma \cap (-\Gamma)$ is finitely generated. Let $\gamma_1,\dots,\gamma_k$ be a set of generators for $\Gamma \cap (-\Gamma)$. Since $\gamma \in \Gamma$ is characterized by $E(\gamma) = \mathbf{1}$, the $\gamma \in \Gamma \cap (-\Gamma)$ is characterized by the fact that $u(\gamma)$ is unitary, and we may then redefine the $u(\gamma)$'s so that

$$(2.9.103)\quad u(n_1\gamma_1 + n_2\gamma_2 + \dots + n_k\gamma_k) = u(\gamma_1)^{n_1}u(\gamma_2)^{n_2} \dots u(\gamma_k)^{n_k}$$

Since $\hat{G}/\Gamma \cap (-\Gamma)$ is finite, there exists a finite set $F \subseteq \hat{G}$ such that

$$(2.9.104)\quad F + (\Gamma \cap (-\Gamma)) = \hat{G}$$

Thus, for any $\gamma \in \hat{G}$, we may arrange the normalization of $u(\gamma)$ so that

(2.9.105) $u(\gamma) = u(\xi)u(\eta)$

where $\xi \in F$, $\eta \in \Gamma \cap (-\Gamma)$ and $\xi + \eta = \gamma$. (If F is minimal with respect to the property 2.9.104, then the decomposition $\gamma = \xi + \eta$ of γ is unique, and if we in addition choose F such that $0 \in F$ and $u(0) = 1$, then 2.9.105 gives a unique definition of $u(\gamma)$ for all $\gamma \in \hat{G}$.) Now, any $x \in M_F^\alpha$ has a Fourier decomposition

(2.9.106) $x = \sum_{\gamma \in \hat{G}} x_\gamma u(\gamma)$

where

(2.9.107) $x_\gamma = \{\int_G dg\, \overline{\gamma(g)}\, \alpha_g(x)\} u(\gamma)^* \in M^\alpha E(\gamma)$

and the sum is finite.
Thus the restriction $\delta|_{M^\alpha}$ has the form

(2.9.108) $\delta(x) = \sum_{\gamma \in K} \delta_\gamma(x) u(\gamma)$

for $x \in M^\alpha$, where K is the finite set given by 2.9.86, and

(2.9.109) $\delta_\gamma(x) = \{\int_G dg\, \overline{\gamma(g)}\, \alpha_g(\delta(x))\} u(\gamma)^*$

for $x \in M^\alpha$, $\gamma \in K$. Then

(2.9.110) $\|\delta_\gamma(x)\| \leq \|\delta|_{M^\alpha}\| \|x\|$.

Since each $u(\gamma)$ is the product of $u(\xi)$'s for ξ in the finite set $F \cup \{\gamma_1, \dots, \gamma_k\}$, it is now clear from Lemma 2.9.4 that the extimates 2.9.95 and 2.9.96 are valid for some $\varepsilon > 0$. Theorem 2.9.17 now follows from Lemma 2.5.4 as in the conclusion of the proof of Theorem 2.9.7.

Example 2.9.21 [BGJ 1, Example 4.14]

We now show that the conclusion of Theorem 2.9.17 fails if G is not assumed to be Lie, even in the strong version of 2.9.77 where $A^\alpha(\gamma)A^\alpha(\gamma)^* = A^\alpha$ for all $\gamma \in \hat{G}$.

There is an (up to isomorphism) unique ergodic action β of the group $\mathbf{Z}_2 \times \mathbf{Z}_2$ on the algebra M_2 of 2×2 matrices. If g_1, g_2 are the generators of $\mathbf{Z}_2 \times \mathbf{Z}_2$, i.e. $g_1^2 = g_2^2 = 1$, $g_1 g_2 = g_2 g_1$, β is defined by

(2.9.111) $\beta_{g_1} = \mathrm{Ad}\begin{pmatrix}1 & 0\\ 0 & -1\end{pmatrix}$, $\beta_{g_2} = \mathrm{Ad}\begin{pmatrix}0 & 1\\ 1 & 0\end{pmatrix}$

Thus we may define a product type action of $G = \overset{\infty}{\times}(\mathbf{Z}_2) \cong \overset{\infty}{\times}(\mathbf{Z}_2 \times \mathbf{Z}_2)$ on the infinite tensor product

(2.9.112) $A = \overset{\infty}{\otimes} M_2$

Then G is compact and abelian, the action α is ergodic so the spectral spaces $A^\alpha(\gamma)$ are one dimensional for $\gamma \in \hat{G} \cong \overset{\infty}{+} \mathbf{Z}_2$, and one easily verifies that each $A^\alpha(\gamma)$ is contained in a finite sub-tensor-product of $\overset{\infty}{\otimes} M_2$ and A_F^α is the *-algebra of finite tensors in $\overset{\infty}{\otimes} M_2$.

The algebra A identifies with the CAR algebra over a separable Hilbert space H, [BR 2], and if $f_1, f_2, \ldots$ is an orthonormal basis for H we may arrange this identification such that A_F^α is the *-algebra generated by the annihilators $a(f)$ where f is in the (non-closed) linear span D of $f_1, \ldots$. Let H be a symmetric operator from D into D with deficiency indices $(0,1)$, and let δ be the corresponding quasifree derivation, determined by

(2.9.113) $\delta(a(f)) = a(iHf)$,

for $f \in D$, see [Bra 4]. Then $\delta(A_F^\alpha) \subseteq A_F^\alpha$. Since H generates a semigroup of isometries which are not unitaries on H, $\bar{\delta}$ generates a group of injective morphisms which are not automorphisms on A. Thus δ is not a pre-generator, and δ does not even have an extension to a generator on A.

By taking an infinite tensor-product of the ergodic action of $\mathbf{T}^2$ on a simple C^*-algebra described in Remark 2.9.2, we obtain correspondingly an ergodic action α of $\mathbf{T}^\infty$ on a simple C^*-algebra A such that not all $\delta \in \mathrm{Der}\,(A_F^\alpha, A_F^\alpha)$ are pregenerators. In this case, considering one-parameter subgroups of $\mathbf{T}^\infty$ with large angular velocity near ∞, one deduces that A_∞ consists of finite tensors, and then that not even all $\delta \in \mathrm{Der}\,(A_\infty, A_\infty)$ are pregenerators.

§2.9.2.3 Compact groups and $(A^\alpha)' \cap A'' = \mathbb{C}1$.

Theorem 2.9.22 (Bratteli-Goodman)

Let α be a faithful action of a compact group G on a C^*-algebra A, and assume there exists a faithful covariant representation π of A such that

(2.9.114) $\pi(A^{\alpha})' \cap \pi(A)'' = \mathbb{C}1$.

Let $\delta \in \mathrm{Der}\,(A_F^{\alpha},A)$ and assume that $\delta(A^{\alpha}) \subseteq A_F^{\alpha}$.
It follows that δ has a unique decomposition

(2.9.115) $\delta = \delta_0 + \tilde{\delta}$

on A_F^{α}, where δ_0 is the generator of a one-parameter subgroup of $\alpha(G)$, and $\tilde{\delta}$ is bounded.

Remark 2.9.23

This is a simplified version of Theorem 2.5 in [BG 1], and we follow the proof there.

The sole purpose of the condition $\delta(A^{\alpha}) \subseteq A_F^{\alpha}$ is to ensure that $\delta|_{A^{\alpha}}$ is implemented by an element in $\pi(A)''$. This condition could therefore be replaced by the condition that $\pi(A^{\alpha})''$ is injective or that $\pi(A)''$ is finite, see [Chr 2, Theorem 2.3]. The theorem may thus well be true without this condition.

Proof of Theorem 2.9.22

We drop the notation π and consider A as acting on the representation Hilbert space H. Put $M = A''$ and let α also denote the extension of α to M. By Theorem 2.3.8 or 2.3.12 δ extends to a derivation from M_F^{α} into M such that if δ also denotes this extension, then $\delta|_{M^{\alpha}(\gamma)}$ is bounded and σ-weakly continuous for each $\gamma \in \hat{G}$.
The assumption $\delta(A^{\alpha}) \subseteq A_F^{\alpha}$ implies by Baires category theorem that $\delta(A^{\alpha}) \subseteq A^{\alpha}(K)$ for a finite subset $K \subseteq \hat{G}$ (see the beginning of the proof of Lemma 2.9.19), and then $\delta(M^{\alpha}) \subseteq M^{\alpha}(K) \subseteq M_F^{\alpha}$. It follows from Lemma 2.9.19 that there exists an $h \in M_F^{\alpha}$ such that

(2.9.116) $\delta(x) = [h,x]$

for $x \in M^{\alpha}$, and we may assume $h^* = -h$. Define derivations δ_0 and $\tilde{\delta}$ from M_F^{α} into M by

(2.9.117) $\tilde{\delta}(x) = [h,x]$, $\delta_0(x) = \delta(x) - \tilde{\delta}(x)$

for $x \in M_F^{\alpha}$. Then

(2.9.118) $\delta_0\big|_{M^\alpha} = 0.$

The condition 2.9.114 and Lemma 2.7.12 implies that for any $\gamma \in \hat{G}$ there exists a skew-symmetric matrix

(2.9.119) $R(\gamma) \in \mathbf{1} \otimes M_{d(\gamma)}$

such that

(2.9.120) $\delta_0(X) = X\,R(\gamma)$

for all $X \in M^\alpha_{d(\gamma)}(\gamma)$. If $x \in M^\alpha(\gamma)$, then

$$x_{ij} = P_{ij}(\gamma)(x) = d(\gamma)\int_G dg\ \overline{\gamma_{ji}(g)}\ \alpha_g(x)$$

is contained in the finite-dimensional linear span $L(x)$ of the orbit $\alpha_G(x)$ of x under α, and we have

$$x = \sum_{i=1}^{d(\gamma)} x_{ii},$$

see Definition 2.2.11. Since $X = [x_{ij}] \in M^\alpha_{d(\gamma)}(\gamma)$, it follows from 2.9.119 and 2.9.120 that

(2.9.121) $\delta_0(x) \in L(x)$

i.e. δ_0 maps the finite-dimensional space $L(x)$ into itself. In particular this means that δ_0 maps all finite-dimensional α-invariant linear subspaces of A^α_F into themselves, so A^α_F consists of entire analytic elements for δ_0. Also, 2.9.120 implies that $\delta_0\big|_{A^\alpha(\gamma)}$ is bounded for each $\gamma \in \hat{G}$, and hence the exponentiated maps $e^{t\delta_0}\big|_{A^\alpha(\gamma)}$ are bounded. It follows from Lemma 2.5.4 as in the conclusion of the proof of Theorem 2.9.7 that $\delta_0\big|_{A^\alpha_F}$ is a pregenerator, and also $\delta_0\big|_{M^\alpha_F}$ is a pregenerator of a σ-weakly continuous one-parameter group β of $*$-automorphisms of M. But as $\delta_0\big|_{M^\alpha} = 0$, we have $\beta_t\big|_{M^\alpha} = 1$, and hence β commutes with the action τ of the unitary group $U(M^\alpha)$ on M by inner automorphisms. But we have

(2.9.122) $M^\tau = M \cap (M^\alpha)' = \mathbb{C}\mathbf{1},$

i.e. the action τ is ergodic. It follows from Araki-Haag-Kastler-Takesaki's duality theorem that $\beta_t \in \alpha_G$ for all t, see Theorem 2.9.32. It follows that δ_0 is the generator of a one-parameter subgroup of α_G. Then $\tilde{\delta} = \delta - \delta_0$ is a bounded derivation of A (actually, as $h \in M_F^\alpha$, we have $\tilde{\delta}(A_F^\alpha) \subseteq A \cap M_F^\alpha = A_F^\alpha$).

This completes the proof of Theorem 2.9.22 except for the uniqueness statement. But if δ_0 is the generator of a one-parameter group of automorphisms in α_G and δ_0 is bounded, then δ_0 extends to M_F^α and $\delta_0|_{M^\alpha} = 0$. By Sakai-Kadisons derivation theorem it exists an $h \in M$ such that $\delta_0 = \mathrm{ad}(h)$, but $h \in (M^\alpha)'$ since $\delta_0|_{M^\alpha} = 0$. Thus $h \in \mathbb{C}\mathbf{1}$ by 2.9.114, and hence $\delta_0 = 0$. This proves uniqueness.

§2.9.3 Applications and examples

By combining the spatial theory of the previous section with results on the existence of covariant representations with specific properties, we can prove decomposition- and generator-properties of derivations in several more or less special situations. In order that these lecture notes shall have a finite length, we will not prove the existence of the particular representations from the dynamical properties, but only state the relevant theorems and give references. There is one particular case, treated in the subsection 2.9.3.2, where we have not been able to separate the spatial and other arguments properly. (See Remark 4.3)

§2.9.3.1 The circle group

Theorem 2.9.24

If $G = \mathbf{T}$ is the circle group, and α is an action of G on a separable C*-algebra A such that all ideals in A are globally fixed under α, then any $\delta \in \mathrm{Der}(A_F^\alpha, A)$ has the decomposition

$$(2.9.123) \qquad \delta = \delta_{\mathrm{inv}} + \tilde{\delta}$$

on A_F^α, where δ_{inv} commutes with A and $\tilde{\delta}$ is bounded. In particular δ is a pregenerator.

This result is an immediate consequence of Theorem 2.9.10 and the following theorem:

Theorem 2.9.25 (Kishimoto)

If $G = \mathbf{T}$ is the circle group, and α is an action of G on a separable C^*-algebra A such that all ideals in A are globally fixed by α, then there exists an α-invariant pure state on A.

Proof

See [Kis 3].

The derivation δ_{inv} in Theorem 2.9.24 is not in general a scalar multiple of the generator δ_0 of α, see Example 2.7.10. However, Theorem 2.6.6 states that δ_{inv} is a bounded perturbation of a scalar multiple of δ_0 in some circumstances, and hence:

Corollary 2.9.26

Let $G = \mathbf{T}$ and let α be an action of G on a separable prime C^*-algebra A with generator δ_0, and assume that $\Gamma(\alpha) \neq \{0\}$ and that all ideals in A are globally fixed by α.

It follows that any derivation $\delta \in \mathrm{Der}\,(A_F^\alpha, A)$ has a (unique) decomposition

$$(2.9.124)\quad \delta = \lambda\delta_0 + \tilde{\delta}$$

on A_F^α, where $\lambda \in \mathbf{R}$ and $\tilde{\delta}$ is bounded.

Remark 2.9.27

A slightly different version of this result is Theorem 3.6 in [BaK 1].

Example 2.7.10 implies that this theorem fails if the condition $\Gamma(\alpha) \neq 0$ is removed, even when A is simple and unital.

Remark 2.9.2 implies that if α is a faithful ergodic action of $G = \mathbf{T}^2$ on a simple C^*-algebra A, then $\Gamma(\alpha) = \hat{\mathbf{T}}^2 \cong \mathbf{Z}^2$, but a derivation $\delta \in \mathrm{Der}\,(A_F^\alpha, A)$ does not necessarily have the decomposition

$$(2.9.125)\quad \delta = \lambda_1\delta_1 + \lambda_2\delta_2 + \tilde{\delta}$$

where $\lambda_1, \lambda_2 \in \mathbf{R}$, δ_1, δ_2 are the generators of the two canonical one-parameter subgroups of $\alpha_{\mathbf{T}^2}$, and $\tilde{\delta}$ is bounded (albeit δ does have this decomposition with $\hat{\delta}$ approximately bounded). Thus the natural generalization of Corollary 2.9.26 to $G = \mathbf{T}^2$ fails. We will however see later, in Theorem 2.9.31, that if one imposes a stronger condition of freeness on α than $\Gamma(\alpha) = \hat{G}$, the decomposition corresponding to 2.9.125 is valid with $\tilde{\delta}$ bounded.

We can prove that any $\delta \in \mathrm{Der}\,(A_F^\alpha, A_F^\alpha)$ is a pre-generator under weaker dynamical assumption than in Theorem 2.9.24:

Theorem 2.9.28 (Bratteli-Kishimoto)

Let α be an action of $G = \mathbf{T}$ on a separable C^*-algebra A, and assume that there exists an $\varepsilon > 0$ such that for each prime ideal P in A, P is either (globally) fixed by α, or $\alpha_t(P) \neq P$ for $0 < t < \varepsilon$.

It follows that all $\delta \in \mathrm{Der}\,(A_F^\alpha, A_F^\alpha)$ are pre-generators.

Remark 2.9.29

This is Theorem 1.3 in [BK 2], and we will follow the proof there.

If $A = C_0(\Omega)$ is abelian with spectrum Ω, $G = \mathbf{T}$ and δ_0 is the generator of α, then it follows from the techniques of Theorem 2.4.20 that all $\delta \in \mathrm{Der}\,(A_F^\alpha, A)$ have the form $\delta = \lambda\delta_0$, where λ is a continuous function on the complement $\Omega \setminus \Omega_0$ of the set of fixed points Ω_0, such that λ is bounded on sets of bounded frequency. The ε-condition in Theorem 2.9.28 implies that there is a maximal finite frequency $1/\varepsilon$, and hence this condition implies that λ is uniformly bounded, and δ is bounded by a multiple of δ_0. This partly explains the significance of the ε-condition. However, if A is abelian the theorem is true without the ε-condition anyway, by restriction to the sub-algebras of functions which are constant on orbits of frequency higher than a given constant, see 2.4.75. See [BK 2, Proposition 4.7].

Proof of Theorem 2.9.28

We will only outline the argument, and refer to Section 4 of [BK 2] for details.

First if α is an action of a compact abelian group G on a C^*-algebra A, and if $x \in A^\alpha(\gamma)$ for some $\gamma \in \hat{G}$, then it follows from Lemma 2.7.7 that x has the form $x = ay$ where $a \in \overline{A^\alpha(\gamma)A^\alpha(\gamma)^*}^{\|\,\|}$ and $y \in A^\alpha(\gamma)$.

It follows that if J is an α-invariant (closed twosided) ideal in A, then $J^{\alpha}J_F^{\alpha} = J_F^{\alpha}$, thus $J_F^{\alpha 2} = J_F^{\alpha}$. Hence, if $\delta : A_F^{\alpha} \to A$ is a derivation it follows from the derivation property that $\delta(J_F^{\alpha}) \subseteq J$, and δ induces a derivation δ_J on J and δ^J on A/J, defined on the G-finite elements of the action induced by α on J and A/J. One then argues that δ is a pregenerator if and only if both δ_J and δ^J are pregenerators, [BK 2, Lemma 4.3]. (A similar result is [GW 1, Lemma 2].)

Returning to the specific situation of Theorem 2.9.28, define

$$(2.9.126)\quad J = \cap\{P \in \mathrm{Prim}(A) \mid \exists t \in \mathbf{T} : \alpha_t(P) \neq P\}$$

where $\mathrm{Prim}(A)$ is the primitive ideal space in A. Note that if A is abelian with spectrum Ω, J is the ideal corresponding to the interior of the set of α-fixed points in G. In general J is α-invariant and any ideal of J is α-invariant. By Theorem 2.9.24, δ_J is a pregenerator (this is the point in the proof where the separability of A is used). On the other hand the derivation δ^J induced by δ on $B = A/J$ is a pregenerator by the following reasoning: Let α also denote the action of G on B induced by α. By the definition 2.9.126 of J we have

$$\cap\,\{P \in \mathrm{Prim}(B) \mid \alpha_t(P) \neq P \text{ for some } t\} = \{0\}$$

Also the ε-assumption of Theorem 2.9.28 implies that if $P \in \mathrm{Prim}(B)$ and $\alpha_t(P) \neq P$ for some t, then $\alpha_t(P) \neq P$ for $0 < t < \varepsilon$. Thus if $p(P)$ is the period of P under α, then $p(P) > \varepsilon$, i.e. $p(P) \in \{1, \frac{1}{2}, \frac{1}{3}, \ldots, \frac{1}{N}\}$, where N is the largest integer such that $\frac{1}{N} \geq \varepsilon$. If $P \in \mathrm{Prim}(B)$ is non-invariant, pick a pure state ω on B with $\ker(\pi_\omega) = P$, where π_ω is the irreducible representation of B defined by ω. The representation

$$\pi_P = \int_{\mathbf{T}}^{\oplus} dt\ \pi_\omega \circ \alpha_t$$

of A on $H_\omega \otimes L^2(\mathbf{T})$ is then α-covariant, and the center of this representation must be a subalgebra of $\mathbf{1} \otimes L^\infty(\mathbf{T})$ which is invariant under rotations by elements in $\mathbf{T}$. Thus the center must be the algebra of periodic L^∞-functions on $\mathbf{T}$ of period p for some $p \in \{1, \frac{1}{2}, \frac{1}{3}, \ldots, 0\}$. But using $\alpha_t(P) \neq P$ for $0 < t < p(P)$ one deduces that $\frac{1}{p}$ divides $\frac{1}{p(P)}$, in particular $p \leq N$. The function $t \to e^{2\pi i \frac{t}{p}}$ then defines a unitary operator u in the center of $\pi_P(A)'' = M$, and $u \in M^{\alpha}(\frac{1}{p})$, where α also denotes the extension of α to M. Then $u^m \in M^{\alpha}(\frac{m}{p})$

for all $m \in \mathbf{Z}$, and thus if M is the least common multiple of the numbers $1,2,\ldots,N$, then $M^\alpha(M)$ contains a unitary operator for all $P \in \mathrm{Prim}(B)$ such that $\alpha_t(P) \neq P$ for some t. Thus, if π is the direct sum of all such π_P, then $(\pi(A)'')^\alpha(M)$ contains a unitary operator. It follows from Theorem 2.9.18 that δ^J is a pregenerator.

We have established that δ_J and δ^J are pregenerators, hence δ is a pregenerator by the first part of the proof. This ends the proof of Theorem 2.9.28.

Problem 2.9.30

Remove the ε-condition from Theorem 2.9.28.

§2.9.3.2 Free actions

We have already remarked that the natural extension of Corollary 2.9.26 from $G = \mathbf{T}$ to $G = \mathbf{T}^2$ is not longer true. However, if one imposes a stronger condition of freeness on α than $\Gamma(\alpha) = \hat{G}$, one obtains:

Theorem 2.9.31 (Bratteli-Kishimoto)

Let G be a second countable locally compact group which is abelian or compact, and α a faithful action of G on a simple separable unital C^*-algebra A. Assume that there exists a sequence $\tau_n \in \mathrm{Aut}\,(A)$ such that

(2.9.127) $\tau_n\alpha_g = \alpha_g\tau_n$

for all $n \in \mathbf{N}$, $g \in G$, and

(2.9.128) $\lim_{n\to\infty} \| \tau_n(x)y - y\tau_n(x)\| = 0$

for all $x,y \in A$. Let $\delta \in \mathrm{Der}\,(A_F^\alpha, A)$ be a derivation, and if G is non-compact assume that $\delta\big|_{A^\alpha(\Omega)}$ is bounded for a compact neighbourhood Ω of 0 in $\hat{G}$.

It follows that δ has a (unique) decomposition

$$\delta = \delta_0 + \tilde{\delta}$$

on A_F^α, where δ_0 is the generator of a one-parameter subgroup of α_G, and $\tilde{\delta}$ is inner.

Remark 2.9.32

This is Theorem 1.1 in [BK 2]. The special case that $[\delta,\tau_n] = 0$ for all n was considered in [KR 3] when G is compact and abelian, in [RST 1] when G is compact and in [BaK 1] when G is locally compact abelian, in all cases without separability conditions on A and G, and the stronger conclusion $\delta = \delta_0$ is then reached. Weaker ergodic properties of the sequence τ_n has been considered in [LP 1] and [BER 2] under the name of topological transitivity and strong topological transitivity, see Definition 2.9.34, but so far results only exist for the case $[\delta,\tau_n] = 0$.

As remarked in the introduction of §2.9.3 we have not been able so separate the spatial argument completely from the other arguments in the proof of Theorem 2.3.31 when G is non-abelian. The extra element needed is a version of the Tannaka duality theorem. For completeness we will state three versions of this theorem which are obviously related. The first version was used in the proof of Theorem 2.9.22. (See Remark 4.3)

Theorem 2.9.33 (Araki-Haag-Kastler-Takesaki)

Let G be a compact group, and α an action of G on a von Neumann algebra M. Let H be another group and τ an action of H on M such that

(2.9.129) $[\alpha,\tau] = 0$

and τ is ergodic in the sense

(2.9.130) $M^{\tau} = \mathbb{C}1.$

Let β be an automorphism of M such that

(2.9.131) $\beta(x) = x$

for all $x \in M^{\alpha}$, and

(2.9.132) $[\beta,\tau] = 0.$

It follows that there exists a $g \in G$ such that

(2.9.133) $\beta = \alpha_g$.

Proof

See [AHKT 1, Theorem III.3.3] and [Tak 1].

For C^*-algebras, the definition 2.9.130 of ergodicity is not a good one, [LP 1], [BER 2]. Instead we use

Definition 2.9.34 ([LP 1], [BER 2])

Let τ be an action of a group H on a C^*-algebra A. We say that τ is topologically transitive if for any pair $x,y \in A \setminus \{0\}$ there exists a $h \in H$ such that

(2.9.134) $x\tau_h(y) \neq 0$

We say that τ is strongly topologically transitive if for any pair $\{x_1,\dots,x_n\}$ and $\{y_1,\dots,y_n\}$ of finite sequences in A of the same length such that

(2.9.135) $\sum_{k=1}^{n} x_k \otimes y_k \neq 0$

there exists a $h \in H$ such that

(2.9.136) $\sum_{k=1}^{n} x_k \tau_n (y_k) \neq 0$

Remark 2.9.35 ([LP 1], [BER 2])

It is easily seen that τ is topologically transitive if and only if any pair of hereditary α-invariant C^*-subalgebras of A has non-trivial intersection, and one immediately deduce the following three facts:

1. Topological transitivity implies ergodicity, but not conversely.
2. If $A = M$ is a von Neumann algebra, topological transitivity is equivalent to ergodicity.
3. If $A = C_0(\Omega)$ is abelian, and T is the H-flow on Ω corresponding to τ, then topological transitivity means that for any pair V_1,V_2

of open subsets of Ω there exists a $h \in H$ such that $V_1 \cap T_h(V_2) \neq \emptyset$, i.e. T is topologically transitive in the sense of ergodic theory.

It is clear that strong topological transitivity implies topological transitivity, but at present it is not clear whether the two consepts are equivalent or not. They are equivalent if A is abelian or H is compact, [BER 2, Theorem 1.3 and Corollary 1.6]. We will need that if A is simple and unital, and τ_n is a sequence of automorphisms satisfying 2.9.127 and 2.9.128, then the group generated by the τ_n's is a strongly topologically transitive group of automorphisms of A, see [KR 3] or [BER 2, Theorem 1.7].

The two other versions of Tannaka's duality theorem are now:

Theorem 2.9.36 (Longo-Peligrad)

Let G be a compact abelian group, and α an action of G on a C^*-algebra A. Let H be another group, and τ an action of H on A such that $[\tau,\alpha] = 0$ and τ is topologically transitive. Let β be an automorphism of A such that $\beta|_{A^\alpha} = 1$ and $[\beta,\tau] = 0$.

It follows that there exists a $g \in G$ such that $\beta = \alpha_g$.

Proof

See [LP 1, Theorem 3.1].

Theorem 2.9.37 (Bratteli-Elliott-Robinson)

Let G be a compact group, and α an action of G on a C^*-algebra A. Let H be another group, and τ an action of H on A such that $[\tau,\alpha] = 0$ and τ is strongly topologically transitive. Let β be an automorphism of A such that $\beta|_{A^\alpha} = 1$ and $[\beta,\tau] = 0$.

It follows that there exists a $g \in G$ such that $\beta = \alpha_g$.

Proof

See [BER 2, Theorem 2.1].

The semi-spatial argument proving Theorem 2.9.31 is based on the existence of a G-covariant representation of A with properties complementary to those of the representation considered in Theorem 2.9.25:

Theorem 2.9.38 (Bratteli-Kishimoto)

Let G be a separable locally compact group which is abelian or compact, and α a faithful action of G on a simple, separable, unital C^*-algebra A. Assume there exists a sequence $\tau_n \in \mathrm{Aut}(A)$ such that $\tau_n\alpha_g = \alpha_g\tau_n$ for all $g \in G$, $n \in \mathbf{N}$, and $\lim_{n\to\infty} \|[\tau_n(x),y]\| = 0$ for all $x,y \in A$.

It follows that there exists a pure state ω on A with corresponding cyclic representation (H_ω, π_ω) such that the center of the direct integral representation

$$(2.9.137) \quad \pi = \int_G^\oplus dg\ \pi_\omega \circ \alpha_g$$

on $H_\omega \otimes L^2(G)$ is the algebra $\mathbf{1} \otimes L^\infty(G)$ of diagonal operators in this representation.

Proof

See [BK 2, Theorem 2.1]. The idea is to use that $\tau_n(x)$ is a central sequence in A for each $x \in A$, and hence any σ-weak limit point of this sequence in any representation is contained in the center of the representation. One uses separability of A and G to find a pure state ω with the property that the set of elements in $\mathbf{1} \otimes L^\infty(G)$ which are such limit points separates points in G.

Proof of Theorem 2.9.31

We follow the proof of Theorem 2.1 in [BK 2].

Let π be the representation described by 2.9.137, and let $M = \pi(A)''$. It follows from the hypotesis on π that the center of M is $\mathbf{1}_\omega \otimes L^\infty(G)$, and then

$$(2.9.138) \quad M = L(H_\omega) \otimes L^\infty(G)$$

If R is the right regular representation of G on $H_\omega \otimes L^2(G)$, i.e.

$$(2.9.139) \quad R_h(\xi \otimes \phi)(g) = \xi\phi(gh)$$

for $\xi \in H_\omega$, $\phi \in L^2(G)$ and $g,h \in G$, then

(2.9.140) $\pi(\alpha_h(x)) = R_h \pi(x) R_h^*$

for all $x \in A$, $h \in G$. It follows that the representation π is G-covariant, and the action α extends to the action of right translation on $M = L(H_\omega) \otimes L^\infty(G)$. We denote the extended action also by α, and we have in particular

(2.9.141) $M^\alpha = L(H_\omega) \otimes \mathbf{1}$.

It follows from Theorem 2.3.12 that $\delta|_{A^\alpha(K)}$ is bounded and σ-weakly continuous in the representation π, and δ extends by σ-weak continuity to $M^\alpha(K)$ for each compact $K \subseteq \hat{G}$. The extended operator, also denoted by δ, on M_F^α, is still a derivation.

Since $L(H_\omega)$ is a type I factor, the restriction of δ to $M^\alpha = L(H_\omega) \otimes \mathbf{1}$ is implemented by a $h = -h^* \in M$. This follows from the reasoning in Example 1.6.4.

We now discuss the case that G is compact, and make use of the notation in §2.2.3. In this case

(2.9.142) $M_F^\alpha = L(H_\omega) \otimes L_F^\infty(G)$

where $L_F^\infty(G)$ is the set of linear combinations of matrix elements of irreducible representations, and the tensor product is algebraic. Let $\gamma \in \hat{G}$ with $d = \dim(\gamma)$. The action α and derivation δ on $L(H_\omega) \otimes L^\infty(G)$ extends to $\overline{\alpha}$ and $\overline{\delta}$ on $L(H_\omega) \otimes L^\infty(G) \otimes M_d$ by tensoring with 1. Then

(2.9.143) $\mathbf{1} \otimes \gamma \in \mathbf{1} \otimes L_F^\infty(G) \otimes M_d$

and if $x \in M_d^\alpha(\gamma)$, then $x \in L(H_\omega) \otimes L_F^\infty(G) \otimes M_d$ and x has the transformation law

(2.9.144) $\alpha_g(x) = x(\mathbf{1} \otimes \mathbf{1} \otimes \gamma(g))$

Thus

(2.9.145) $x(\mathbf{1} \otimes \gamma^*) \in (M \otimes M_d)^\alpha = N \otimes \mathbf{1} \otimes M_d$

Hence

(2.9.146) $\overline{\delta}(x(\mathbf{1}\otimes\gamma^*)) = [h\otimes\mathbf{1}_d, x(\mathbf{1}\otimes\gamma^*)]$

and thus

(2.9.147) $\overline{\delta}(x)(\mathbf{1}\otimes\gamma^*) + x\overline{\delta}(\mathbf{1}\otimes\gamma^*) = [h\otimes\mathbf{1}_d, x](\mathbf{1}\otimes\gamma^*)$

since $h\otimes\mathbf{1}_d$ commute with $\mathbf{1}\otimes\gamma^*$. Multiplying to the right with $\mathbf{1}\otimes\gamma$ we obtain

(2.9.148) $\overline{\delta}(x) = [h\otimes\mathbf{1}_d, x] - x\overline{\delta}(\mathbf{1}\otimes\gamma^*)(\mathbf{1}\otimes\gamma)$.

But

(2.9.149) $\overline{\delta}(\mathbf{1}\otimes\gamma^*)(\mathbf{1}\otimes\gamma) \in \mathbf{1}\otimes L^\infty(G)\otimes M_d$

since $\delta(\text{Centre}(M)\cap D(\delta)) \subseteq \text{Centre}\ M$, by the derivation property of δ.

In the expression 2.9.148 for $\overline{\delta}(x)$, all elements are contained in $L(H_\omega)\otimes L^\infty(G)\otimes M_d$, and can therefore be viewed as functions from G into $L(H_\omega)\otimes M_d$. If we evaluate these functions at g for an $x\in A_d^\alpha(\gamma)$, and supress the π, we obtain

(2.9.150) $\alpha_g(\delta(x)) = [h(g)\otimes\mathbf{1}_d, \alpha_g(x)] - \alpha_g(x)(\delta(\mathbf{1}\otimes\gamma^*)(\mathbf{1}\otimes\gamma^*))(g)$

for almost all $g\in G$. Since A and G are second countable, and thus $\hat{G}$ is countable, there is a $g\in G$ such that 2.9.150 is valid for all $x\in A_{d(\gamma)}^\alpha(\gamma)$ and all $\gamma\in\hat{G}$. But by 2.9.149, $\delta(\mathbf{1}\otimes\gamma^*)(\mathbf{1}\otimes\gamma^*)(g)$ is a scalar matrix in M_d, and hence

(2.9.151) $[h(g)\otimes\mathbf{1}_d, \alpha_g(x)] \in A\otimes M_d$

for all $x\in A_{d(\gamma)}^\alpha(\gamma)$ and $\gamma\in\hat{G}$. It follows that

$$[h(g), A_F^\alpha] \subseteq A$$

and $\text{ad}(h(g)) \equiv \tilde{\delta}'$ defines a bounded derivation of A. Thus $\tilde{\delta} = \alpha_g^{-1}\tilde{\delta}'\alpha_g$ is a bounded derivation of A. If we define $\delta_0 = \delta - \tilde{\delta}$, then $\delta_0\in \text{Der}(A_F^\alpha, A)$, $\delta_0\big|_{A^\alpha} = 0$ and

(2.9.152) $\delta_0(x) = -x(\delta(1 \otimes \gamma^*)(1 \otimes \gamma))(g) = xR(\gamma)$

for $x \in A^{\alpha}_{d(\gamma)}(\gamma)$, where $R(\gamma)$ is a scalar matrix.

Now, using the argument following formula 2.9.120 in the proof of Theorem 2.3.21, it follows from 2.9.152 that δ_0 maps each finite-dimensional G-invariant subspace of A into itself, and δ_0 is a pregenerator. But the assumption $[\tau_n,\alpha] = 0$ implies that $A^{\alpha}_{d(\gamma)}(\gamma)$ is τ_n-invariant, and hence 2.9.152 implies that δ_0, and then $e^{t\overline{\delta_0}}$, commutes with τ_n for all n. But the group generated by τ_n is strongly topologically transitive by Remark 2.9.35, and it follows from Theorem 2.9.37 that $e^{t\overline{\delta_0}} \in \alpha_G$ for all $t \in \mathbf{R}$. Since $\tilde{\delta} = \delta - \delta_0$ is bounded and A is simple and unital, δ_0 is inner by Sakai's theorem, and this proves the existence of the decomposition in Theorem 2.9.31 when G is compact. (See Remark 4.3)

When G is non-compact, but abelian we deduce the formula corresponding to 2.9.148,

(2.9.153) $\overline{\delta}(x) = [h,x] - x\overline{\delta}(\overline{\gamma})\gamma$

when $x \in M^{\alpha}(\gamma)$, $\gamma \in \hat{G}$, as before. This time we cannot use Tannaka's Duality theorem in the form of Theorem 2.9.37, but since the map $\gamma \in \hat{G} \to \overline{\delta}(\overline{\gamma})\gamma \in 1 \otimes L^{\infty}(G)$ is an additive character, it is possible to use Pontryagin's duality theorem directly, together with a measure theoretic argument which is more complicated than in the previous case due to the fact that $\hat{G}$ is now not discrete, to deduce the existence of the decomposition in Theorem 2.9.31. See [BK 2] for details.

Finally, we note that the decomposition is unique both when G is compact and abelian since the existence of the representation π in Theorem 2.9.38 prevents all the non-zero generators δ_0 of one-parameter subgroups of α_G from being inner. This ends the proof of Theorem 2.9.31.

Remark 2.9.39

Note that the proof of Theorem 2.9.31 only depends on the existence of the representation π with the properties of Theorem 2.9.38, and not on the existence of τ_n, up to the point where we deduce that δ_0 is a pregenerator of a one-parameter group β of automorphisms of A such that each α-invariant closed subspace of A is β invariant. But this latter property of β does not suffice to prove that $\beta_t \in \alpha_G$

for each t, unless G is abelian, see [BER 2, Examples 2.3 and 2.4].

Another example is the following, [A. Kishimoto, private communication]: Let α be the action on $A = \bigotimes_{n=1}^{\infty} M_2$ defined in Example 2.6.7, but choose λ_n to be an increasing sequence of positive integers increasing fast enough that any integer m can be represented as $m = \sum_n \varepsilon_n \lambda_n$ in at most one way where $\varepsilon_n \in \{-1,0,1\}$. Then α is a representation of $\mathbf{T}$, and A^{α} is the tensor product of the canonical diagonal algebras $\begin{pmatrix} \mathbb{C} & 0 \\ 0 & \mathbb{C} \end{pmatrix}$ in M_2, and each $A^{\alpha}(\gamma)$ has the form $A^{\alpha}u(\gamma)$ where $u(\gamma)$ is a single partial isometry. Thus each closed α_t-invariant subspace of A is the closure of its intersection with the algebra of finite tensors in $\bigotimes_{n=1}^{\infty} M_2$. One easily deduces that any infinite tensor product action $t \in \mathbf{R} \to \beta_t$ of the form described in Example 2.6.7 leaves each closed α-invariant subspace invariant. But $\beta_t \neq \alpha_{\mathbf{R}}$ for $t \neq 0$ if the sequence (μ_n) is chosen appropriately.

§2.9.3.3 Compact abelian Lie groups

The decomposition $\delta = \delta_0 + \tilde{\delta}$ in Corollary 2.9.26 and Theorem 2.9.31 immediately implies that δ is a pregenerator. Under weaker assumptions of freeness on α one can conclude that δ is a generator without having a proof of the decomposition:

Theorem 2.9.40 (Bratteli-Kishimoto)

Let G be a compact abelian Lie group (i.e. $G = \mathbf{T}^d \times F$ where d is a natural number and F is a finite abelian group), and let α be an action of G on a C^*-algebra A such that $\Gamma(\alpha) = \hat{G}$ and A is G-prime (Definition 2.6.3). Let $\delta \in \mathrm{Der}(A_F^{\alpha}, A_F^{\alpha})$.

It follows that δ is a pregenerator. Furthermore, there exists an $\varepsilon > 0$ such that

$$(2.9.154) \qquad \sum_{n=0}^{\infty} \frac{\varepsilon^n}{n!} \|\delta^n(x)\| < +\infty$$

for all $x \in A_F^{\alpha}$.

This is Theorem 1.2 in [BK 2], and it is an immediate corollary of Theorem 2.9.18 and the following result:

Theorem 2.9.41 (Bratteli-Kishimoto)

Let G be a compact abelian separable group, and α an action of G on a C^*-algebra A such that $\Gamma(\alpha) = \hat{G}$ and A is G-prime.

It follows that:

(1) There exists a pure state ϕ on A^α such that

$$\left\| \phi \Big|_{A^\alpha(\gamma_1)\ldots A^\alpha(\gamma_n)A^\alpha(\gamma_n)^*\ldots A^\alpha(\gamma_1)^*} \right\| = 1 \tag{2.9.155}$$

for all finite sequences $\gamma_1,\ldots,\gamma_n \in \hat{G}$.

(2) If ϕ is a state satisfying 2.9.155 and $\omega = \phi \circ P_0$, where $P_0 = \int_G dg\, \alpha_g$, then

$$[\pi_\omega(A^\alpha(\gamma))H_\omega] = H_\omega \tag{2.9.156}$$

for all $\gamma \in \hat{G}$, where (H_ω, π_ω) is the cyclic representation defined by ω.

Proof

See [BK 2, Theorem 3.1]. The starting point of the proof is that A^α is prime by [Ped 1, Theorem 8.10.4], and it follows by induction on n that all the ideals $A^\alpha(\gamma_1)\ldots A^\alpha(\gamma_n)A^\alpha(\gamma_n)^*\ldots A^\alpha(\gamma_1)^*$ in A^α are non-zero.

§2.9.3.4 Some special examples

We first consider quasifree actions on the CAR-algebra. Let h be a Hilbert space. The CAR-algebra $A(h)$ over h is generated as a C^*-algebra by the range of an antilinear map

$$f \in h \to a(f) \in A(h) \tag{2.9.157}$$

satisfying

$$\begin{aligned} a(f)a(g)^* + a(g)^*a(f) &= \langle f,g \rangle 1 \\ a(f)a(g) + a(g)\,a(f) &= 0 \end{aligned} \tag{2.9.158}$$

for all $f,g \in h$. The algebra A is uniquely determined by these relations, and as a consequence any unitary operator U on h defines an automorphism α_U on A by

(2.9.159) $\alpha_U(a(f)) = a(Uf)$,

see [BR 2, Theorem 5.2.5]. The automorphism α_U is said to be quasi-free.

Theorem 2.9.42 (Kishimoto)

Let A be the CAR-algebra and let α be an action of $\mathbf{R}$ on A as a group of quasi-free automorphisms. Let $\delta \in \mathrm{Der}(A_\infty, A)$.

It follows that δ has a decomposition

(2.9.160) $\delta = \delta_{inv} + \tilde{\delta}$

on A_∞, where δ_{inv} commutes with α and $\tilde{\delta}$ is bounded. In particular δ is a generator. If furthermore $\Gamma(\alpha) \neq \{0\}$, we can take

(2.9.16) $\delta_{inv} = \lambda\delta_0$

where $\lambda \in \mathbf{R}$ and δ_0 is the generator of α.

Proof

This is the Corollary in [Kis 2], and we follow the proof there.

First note that A is simple, and the Fock vacuum state is pure and α-invariant, [BR 2], hence Theorem 2.9.10 applies to show that δ is a pre-generator (the condition $\| \delta|_{A^\alpha(K)} \| < +\infty$ is fulfilled by the initial remarks to §2.9). Let $t \in \mathbf{R} \to U_t$ be the one-parameter unitary group on h corresponding to α by 2.9.159, and let H be the self adjoint generator of U. If H has pure point spectrum, then α is almost periodic, and thus

(2.9.162) $\delta_{inv}(x) = \lim_{T\to\infty} \frac{1}{2T} \int_{-T}^{T} dt\, \alpha_t(\delta(\alpha_{-t}(x)))$

exists for all $x \in A_\infty$, and δ_{inv} commutes with α. But $\delta - \delta_{inv}$ is bounded by the estimate 2.9.67 in Theorem 2.9.10. If in addition $\Gamma(\alpha) \neq \{0\}$, then $\delta_{inv} = \lambda\delta_0 + \delta'$ where δ' is bounded by Theorem 2.6.6.

If H has a continous part, a KMS state of the reduced system gives a type III_1 factor representation with $\Gamma(\bar{\alpha}) = \hat{\mathbf{R}}$, where $\bar{\alpha}$ denotes the extension of $\bar{\alpha}$ to the weak closure.

The conclusion $\delta = \lambda\delta_0 + \tilde{\delta}$ now follows from Corollary 2.9.17.

In particular theorem 2.9.42 implies to the gauge action, i.e. the action of $\mathbf{T}$ defined by

$$(2.9.163)\quad \alpha_t(a(f)) = a(e^{it}f)$$

for $t \in \mathbf{R}$, $f \in h$. Using the identification of $A(h)$ with the infinite tensor product $\overset{\infty}{\otimes} M_2$, this action identifies with $\overset{\infty}{\otimes} \mathrm{Ad}\begin{pmatrix} 1 & 0 \\ 0 & e^{it} \end{pmatrix}$. More generally, we have

<u>Theorem 2.9.43</u> (Bratteli-Goodman)

Let M_n be the C^*-algebra of $n \times n$ matrices, let β be an action of a compact group G on M_n and let $\alpha = \overset{\infty}{\otimes} \beta$ be the corresponding infinite tensor product action of G on $A = \overset{\infty}{\otimes} M_n$. Let $\delta \in \mathrm{Der}(A_F^\alpha, A)$.

It follows that δ has a unique decomposition

$$(2.9.164)\quad \delta = \delta_0 + \tilde{\delta}$$

on A_F^α, where δ_0 generates a one-parameter subgroup of α_G, and $\tilde{\delta}$ is inner.

<u>Remark 2.9.44</u>

This is Theorem 2.1 in [BG 1]. A pioneering result in this direction is in [PP 1]: If $G = \mathrm{Aut}(M_n)$ and β is the canonical action of G on M_n, and $\delta : D(\delta) \subseteq A \to A$ is a derivation such that $A^\alpha \subseteq D(\delta)$, $\delta\big|_{A^\alpha} = 0$ and $\omega \circ \delta = 0$ where ω is the unique trace state on A, then δ extends to the generator of a one-parameter subgroup of α_G.

We may assume that α is faithful, and then G is a closed subgroup of $\mathrm{Aut}(M_n)$, so G is Lie.

The theorem does not extend to unrestricted product type actions

$$(2.9.165)\quad \prod_m G_m \ni (g_m) \to \bigotimes_m \beta_m(g_m),$$

see Example 2.9.21.

<u>Proof of Theorem 2.9.43</u>

If π is the cyclic representation defined by the trace state ω on A, then π is α-covariant, and an argument using that A^α contains canonical unitaries implementing all finite permutation automorphisms of $A = \overset{\infty}{\otimes} M_n$ shows that

$$\pi(A^\alpha)' \cap \pi(A)'' = \mathbb{C}\,\mathbf{1}\,, \tag{2.9.166}$$

see [PP 1] or [BJ 2, Lemma 4.2]. The theorem is thus a consequence of Theorem 2.9.22 and Remark 2.9.23. More directly, the theorem is a consequence of Theorem 2.9.31.

Next we consider derivations on the Cuntz's algebras $\mathcal{O}_n$, [Cun 1]. This is a unital algebra generated by n operators $s_1,\dots,s_n$ satisfying

$$s_k^* s_k = \mathbf{1}, \quad \sum_{k=1}^{n} s_k s_k^* = \mathbf{1}. \tag{2.9.167}$$

These relation determine $\mathcal{O}_n$ up to canonical isomorphism, thus $\mathcal{O}_n$ is simple. If $U = [u_{ij}]$ is a unitary n×n matrix and one define

$$s_k' = \sum_j u_{jk} s_j \tag{2.9.168}$$

then the s_k' also satisfy 2.9.167, and by canonical uniqueness of $\mathcal{O}_n$ the unitary matrix determines an automorphism α_u of $\mathcal{O}_n$ through

$$\alpha_u(s_k) = s_k' = \sum_j u_{jk} s_j \tag{2.9.169}$$

Then $U \in U(n) \to \alpha_u \in \mathrm{Aut}(\mathcal{O}_n)$ is a representation of $U(n)$ in the automorphism group of $\mathcal{O}_n$.

Theorem 2.9.45 (Bratteli-Goodman)

Let $\mathcal{O}_n$ be the Cuntz algebra, let G be a closed subgroup of U(n) and let α be the restriction to G of the canonical action α of U(n) on $\mathcal{O}_n$. Let $\delta \in \mathrm{Der}((\mathcal{O}_n)^\alpha_F, \mathcal{O}_n)$.

It follows that δ has a unique decomposition

$$\delta = \delta_0 + \tilde{\delta} \tag{2.9.170}$$

on $(\mathcal{O}_n)^\alpha_F$, where δ_0 generates a one-parameter subgroup of α_G and $\tilde{\delta}$ is inner.

Remark 2.9.46

This is Theorem 2.4 in [BG 1].

It is possible to argue that the decomposition 2.9.170 is also unique with respect to the property that $\tilde{\delta}$ is approximately inner, i.e. any nonzero generator δ_0 of a one-parameter subgroup of α_G is not approximately inner, see [BG 1, page 169-170]. A remarkable corollary is that if a derivation $\delta \in \mathrm{Der}((\mathcal{O}_n)^{\alpha}_F, \mathcal{O}_n)$ is approximately inner, it is actually inner.

It is possible to give an argument using Lemma 2.9.4 to show that any derivation δ mapping the *algebra $P(\mathcal{O}_n)$ of polynomials in $s_1,\dots,s_n,s_1^*,\dots,s_n^*$ into itself has all elements in $P(\mathcal{O}_n)$ as analytic elements, and then it is a generator because any $\delta \in \mathrm{Der}(P(\mathcal{O}_n),\mathcal{O}_n)$ is conservative, see [BEv 2, Corollary 2.6 and Remark 3.6].

There exists derivations $\delta \in \mathrm{Der}(P(\mathcal{O}_n),\mathcal{O}_n)$ which are not approximately inner, [BEGJ 1, Example 5.1]. It is known that if $\delta \in \mathrm{Der}(P(\mathcal{O}_n),P(\mathcal{O}_n))$ and if δ commutes with the gauge group, i.e. with the automorphisms α_t of $\mathcal{O}_n$ determined by $\alpha_t(s_k) = e^{it}s_k$, $k=1,\dots,n$, $t \in \mathbf{R}$, then δ is inner if it is approximately inner, [BEGJ 1, Theorem 4.1]. It is an open problem in general whether all approximately inner derivations in $\mathrm{Der}(P(\mathcal{O}_n),P(\mathcal{O}_n))$ are inner.

Proof of Theorem 2.9.45

We outline the argument from [BG 1]. The fixed point algebra F of $\mathcal{O}_n$ under the gauge action τ is canonically isomorphic to the infinite tensor product $\overset{\infty}{\otimes} M_n$, [Cun 1], and the restriction of canonical representation α of $U(n)$ to F identifies with $\overset{\infty}{\otimes} \mathrm{Ad}(\cdot)$. The algebra F has a unique trace state ϕ. If $P_0 = \int_{\mathbf{T}} dt\,\tau_t$, ϕ defines an α-invariant state ω on A by $\omega = \phi \circ P_0$. If π is the corresponding representation of $\mathcal{O}_n$, an elaboration of the argument used in the proof of Theorem 2.9.42 shows that

$$\pi(F)' \cap \pi(\mathcal{O}_n)'' = \mathbb{C}\,1$$

see [BG 1, Lemma 2.2], [BEv 2, Theorem 3.2], [DR 1]. Theorem 2.9.45 is now a consequence of Theorem 2.9.22.

§2.10 Almost invariant non-commutative vectorfields

In §2.9 we considered general derivations defined on the smooth elements for special group actions, and in §2.7 we considered generator properties for invariant derivations defined on the smooth elements for general group actions. In this section we will mention two results of an intermediate nature: There are no restrictions on the actions (the theorems are actually corollaries of more general Banach space results), but the derivation is assumed to have certain approximate commutation properties with the action. Since the proofs of these results are long and based on completely different methods from the ones we have considered so far, we will omit the details.

Theorem 2.10.1 (Goodman-Jørgensen)

Let G be a Lie group, and α an action of G on a C^*-algebra A. Let $\delta \in \mathrm{Der}(A_1,A)$. Assume that the linear span of the set of translations $\alpha_g \circ \delta \circ \alpha_{g^{-1}}$, where g runs through the connected component G_0 of e in G, is finite-dimensional.

It follows that δ is a pregenerator. Furthermore, $A_\infty \subseteq \bigcap_{k=1}^{\infty} D(\delta^k)$, every analytic vector for α is also analytic δ, and the set of analytic vectors for α is a core for δ.

This is Theorem 4.1 in [GJ 2]. Note that the assumption $D(\delta) = A_1$ implies that δ is conservative, Theorem 2.3.6, and hence the generator property of δ follows once the fact that α-analytic vectors are analytic for δ is shown. Note also that if G is abelian or compact, the condition on δ implies that $\delta(A_F^\alpha) \subseteq A_F^\alpha$, and thus the theorem corrobrates the conjecture indicated in the first sentence in §2.9.1.2.

The next theorem extends Theorem 2.10.1 in the case $G = \mathbf{R}$.

Theorem 2.10.2 (Robinson)

Let σ be a strongly continuous one-parameter group of isometries on a Banach space B with generator δ, and let H be an operator from B_∞ into B such that

(2.10.1) H is dissipative

and there is a constant $C > 0$ such that

(2.10.2) $\|[\sigma_t,H]\,x\| \leq C|t|(\|x\| + \|\delta(x)\|)$

It follows that the closure of H is the generator of a strongly continuous semigroups of contractions.

This is Corollary 1.2 in [Rob 2]; a preliminary version was proved in [BaR 3]. It is not hard to extend the theorem from $G = \mathbf{R}$ to $G = \mathbf{R}^d$, but the general Lie group case is still open.

The theorem fails if the estimate 2.10.2 is replaced by

$$(2.10.3) \quad \| [\sigma_t, H] x \| \leq C|t| (\|x\| + \|\delta(x)\| + \|\delta^2(x)\|),$$

see [Rob 1, Section 4].

Note that if H is a dissipative operator such that $H(\mathcal{B}_\infty) \subset \mathcal{B}_1$, then the uniform boundedness principle implies that there is a $C > 0$ and a n such that:

$$(2.10.4) \quad \| [\sigma_t, H] x \| \leq C|t| \left(\sum_{k=0}^{n} \| \delta^k x \| \right)$$

In this sense Theorem 2.10.2 could be viewed as a step in the direction of solving Problem 2.1.4 in general. But to overcome the problems connected with the counter-example mentioned above and the removal of 2.10.1, explicit operator-algebraic methods seems to be necessary. (See Remark 4.5)

Chapter 3. Dissipations

Recall from Definition 1.1.1. that if A is a C^*-algebra, and Δ is a *-linear map defined from a dense *-subalgebra $D(\Delta)$ of A into A, then Δ is called a dissipation if

$$\Delta(x^*x) \leq \Delta(x^*)x + x^*\Delta(x)$$

for all $x \in D(\Delta)$, and Δ is called a complete dissipation if

$$[\Delta(x_i^*x_j)] \leq [\Delta(x_i^*)x_j + x_i^*\Delta(x_j)]$$

for all finite sequences $x_1,\ldots,x_n \in A$. If G is a locally compact group and α an action of G on A, we will again consider the problems of classifying and proving generator properties of dissipations and complete dissipations mapping one class of smooth elements into another. However, the theory for dissipations are much more rudimentary than for derivations. Consider for example the ultraspecial case that G is a compact abelian Lie group, α is ergodic and A is simple. We established in Theorem 2.9.1 that all $\delta \in \mathrm{Der}(A_F^\alpha, A_F^\alpha)$ has a decomposition

$$\delta = \delta_0 + \tilde{\delta}$$

where δ_0 is the generator of a oneparameter subgroup of α_G and $\tilde{\delta}$ is a bounded derivation. To my knowledge, no analogous result has been established for $\Delta \in \mathrm{Diss}(A_F^\alpha, A_F^\alpha)$, and it is not even clear what the formulation should be. The derivation δ_0 above is the invariant part of δ, and a first step in proving an analogous result for dissipations is thus to characterize the invariant part

$$\Delta_0 = \int_G dg\ \alpha_g \circ \Delta \circ \alpha_{-g}$$

of Δ. This is done in Corollary 3.3.2 . It turns out that Δ_0 is a sum of three terms. 1. An invariant derivation (corresponding to δ_0); 2. An invariant elliptic homogeneous second order differential operator in the generators of the one-parameter subgroups of α_G, and 3. a superposition of the maps $-\alpha_g$ over $g \in G$. In §3.3 we will more generally study invariant dissipations Δ on A_F^α such that $\Delta\big|_{A^\alpha} = 0$ for a compact abelian group G.

In §3.2 we will study the rather satisfying theory for everywhere defined complete dissipations, and in §3.4 some partial results for local dissipations and abelian C^*-algebras are given.

§3.1 Bounded dissipations

Theorem 3.1.1 (Kishimoto)

An everywhere defined dissipation Δ is bounded, and $-\Delta$ is dissipative.

Proof

If the C^*-algebra A has a unit, this is a direct consequence of Proposition 1.4.7 and Theorem 1.4.9. If A does not have a unit, extend the dissipation Δ to the unit extension $A + \mathbb{C}\mathbf{1}$ by putting $\Delta(\mathbf{1}) = 0$. Then

$$
\begin{aligned}
&\Delta((\lambda\mathbf{1} + x)^*(\lambda\mathbf{1} + x)) \\
&= \bar{\lambda}\Delta(x) + \lambda\Delta(x^*) + \Delta(x^*x) \\
&\leq \lambda\Delta(x) + \lambda\Delta(x^*) + \Delta(x^*)x + x^*\Delta(x) \\
&= \Delta((\lambda\mathbf{1} + x)^*)(\lambda\mathbf{1} + x) + (\lambda\mathbf{1} + x)^*\Delta(\lambda\mathbf{1} + x)
\end{aligned} \tag{3.1.1}
$$

for $\lambda \in \mathbb{C}$, $x \in A$, so the extended Δ is a dissipation and the previous argument applies.

Theorem 3.1.2 (Evans - Hanche-Olsen)

Let Δ be a bounded *-linear map on a C^*-algebra A and define

$$e^{-t\Delta} = \sum_{n=0}^{\infty} \frac{(-t)^n}{n!} \Delta^n \tag{3.1.2}$$

for $t \geq 0$.

The following conditions are equivalent:

$$\Delta(x^*x) \leq \Delta(x^*)x + x^*\Delta(x) \tag{3.1.3}$$

for all $x \in A$, and

$$e^{-t\Delta}(x^*x) \geq e^{-t\Delta}(x^*)e^{-t\Delta}(x) \tag{3.1.4}$$

for all $x \in A$, $t \geq 0$.

Proof

This is Corollary 3 in [EHO 1].

The implication 3.1.4 ⇒ 3.1.3 is immediate by differentiation at $t = 0$, so we consider the other implication.

Furstly, we show that 3.1.3 implies that $e^{-t\Delta}$ is positivity preserving for $t > 0$.

If $y \in A$ and $y \geq 0$, and ω is state with $\omega(y) = 0$, we argue that $\omega(\Delta(y)) \leq 0$. By Schwarz inequality we have $\omega(y^{\frac{1}{2}}z) = 0 = \omega(zy^{\frac{1}{2}})$ for all $z \in A$. Thus the inequality $\Delta(y) \leq y^{\frac{1}{2}}\Delta(y^{\frac{1}{2}}) + \Delta(y^{\frac{1}{2}})y^{\frac{1}{2}}$ implies that $\omega(\Delta(y)) \leq 0$.

Next, if $y \in A$ and $y \geq 0$, and $ya = 0$ for an $a \in A$, then $a^*\Delta(y)a \leq 0$. This is because the hypothesis implies that $\omega(a^*ya) = 0$ for all states ω, then $\omega(a^*\Delta(y)a) \leq 0$ by the previous paragraph, and thus $a^*\Delta(y)a \leq 0$.

The resolvent $(1 + \lambda\Delta)^{-1}$ exists for small positive λ since Δ is bounded. We now argue that $(1 + \lambda\Delta)^{-1}$ is positivity preserving. It is enough to show that if $x = x^* \in A$ satisfies $(1 + \lambda\Delta)(x) \geq 0$, then $x \geq 0$. Let $x = x_+ - x_-$ be the decomposition of x in a positive and negative part. Then $x_+x_- = 0$ and, by the previous paragraph, $x_-\Delta(x_+)x_- \leq 0$. Furthermore

$$
\begin{aligned}
& 0 \leq x_-(1 + \lambda\Delta)(x)x_- \\
(3.1.5) \qquad & \quad = x_-xx_- + \lambda x_-\Delta(x)x_- \\
& \quad = -(x_-)^3 + \lambda x_-\Delta(x_+)x_- - \lambda x_-\Delta(x_-)x_-
\end{aligned}
$$

Hence

$$(3.1.6) \qquad 0 \leq (x_-)^3 \leq -\lambda x_-\Delta(x_-)x_-$$

from which we deduce

$$(3.1.7) \qquad \|x_-\|^3 \leq \|\Delta\|\|x_-\|^3$$

Choosing $\lambda < \|\Delta\|^{-1}$, it follows that $x_- = 0$ and hence $(1 + \lambda\Delta)^{-1}$ is positivity preserving.

The relation

$$(3.1.8)\qquad e^{-t\Delta} = \lim_{n\to\infty}\left(1+\frac{t}{n}\Delta\right)^{-n}$$

now implies that $e^{-t\Delta}$ is positivity preserving for all $t > 0$, see 1.5.2.

Note in passing that if A has a unit $\mathbf{1}$ and $\Delta(\mathbf{1}) = 0$, the previous argument can be simplified as follows: It follows from Theorem 3.1.1 and Theorem 1.5.2 that $e^{-t\Delta}$ is a semigroup of contractions on A, and as $e^{-t\Delta}(\mathbf{1}) = \mathbf{1}$ this group is positivity preserving. Even in the general case, whether A has a unit or not, A may be embedded as an ideal of codimension 1 is a unital C^*-algebra, and one may employ the trick used in the proof of Theorem 3.1.1 to reduce the proof to the case that $\mathbf{1} \in A$ and $\Delta(1) = 0$.

Once it has been established that $e^{-t\Delta}$ is positivity preserving, the generalized Schwarz inequality 3.1.4 follows from the dissipation inequality and the calculation:

$$\begin{aligned}
&e^{-t\Delta}(x^*x) - e^{-t\Delta}(x^*)e^{-t\Delta}(x)\\
&= \int_0^t ds \frac{d}{ds}\, e^{-s\Delta}(e^{-(t-s)\Delta}(x^*)e^{-(t-s)\Delta}(x))\\
(3.1.9)\qquad &= \int_0^t ds\; e^{-s\Delta}\Big(-\Delta(e^{-(t-s)}(x)\; e^{-(t-s)}(x))\\
&\quad + \Delta(e^{-(t-s)\Delta}(x^*))e^{-(t-s)\Delta}(x) + e^{-(t-s)\Delta}(x^*)\Delta(e^{-(t-s)\Delta}(x))\Big)\\
&\geq 0,
\end{aligned}$$

see [Eva 1]. This ends the proof of Theorem 3.1.2.

Remark 3.1.3

In the course of the proof of Theorem 3.1.2 it was established that if Δ is a bounded *-linear map on a C^*-algebra with the property

$$(3.1.10)\qquad \Delta(x^2) \leq \Delta(x)x + x\Delta(x)$$

for all $x = x^* \in A$, then $-\Delta$ is the generator of a positive semigroup on A. Actually, if A is unital, then $-\Delta$ is the generator of a positive semigroup if and only if it satisfies any of the following equivalent conditions

(3.1.11) $\begin{cases} \text{If } x \in A,\ x \geq 0 \text{ and } \omega \text{ is a state with } \omega(x) = 0, \text{ then} \\ \omega(\Delta(x)) \leq 0, \end{cases}$

(3.1.12) $\Delta(x^2) - x\Delta(\mathbb{1})x \leq \Delta(x)x + x\Delta(x)$ for all $x = x^* \in A$,

(3.1.13) $\Delta(\mathbb{1}) - u\,\Delta(\mathbb{1})u \leq \Delta(u^*)u + u^*\Delta(u)$ for all unitaries $u \in A$,

see [EHO 1, Theorem 2]. For some extensions to unbounded Δ, see [BR 3]. For an extensive abstract study of generators of positive semigroups, see [BaR 1].

§3.2 Bounded complete dissipations

To classify general bounded dissipations on a non-abelian C^*-algebra is as herculean a task as to classify general positive maps, and the situation appears gloomy even for low-dimensional matrix algebras, [Cho 1], [Cho 2]. However, the situation for complete dissipations is much brighter, and we will see that Sakai-Kadison's derivation theorem has an extension to complete dissipations, see Remark 2.5.12 and Theorem 3.2.2 . This corresponds to the situation for completely positive maps. Recall that a map $\phi : A \to B$, where A, B are C^*-algebras is called completely positive if the induced maps $\phi \otimes 1_n : A \otimes M_n \to B \otimes M_n$ are positive for each full matrix algebra M_n, $n = 1,2,\ldots$. If $B \subseteq L(H)$, then ϕ is completely positive if and only if ϕ has the form

(3.2.1) $$\phi(x) = V^*\pi(x)V$$

for all $x \in A$, where π is a representation of A on a Hilbert space K, and V is a bounded linear map from H into K, [Sti 1]. This is the so called Stinespring representation of completely positive maps.

We say that a *-linear map $\phi : A \to B$ is n-positive if $\phi \otimes 1_n : A \otimes M_n \to B \otimes M_n$ is positive, and ϕ is said to be strongly n-positive if the matrix inequality

(3.2.2) $$[\phi(x_i^* x_j)] \geq [\phi(x_i)^*\phi(x_j)]$$

is valid for all sequences $x_1, x_2, \ldots, x_n$ of length n in A. It can be proved, [Eva 1], that if ϕ is a n-positive contraction then ϕ is strongly n-1-positive for $n = 2,3,\ldots$, and if ϕ is strongly n-positive, then ϕ is n-positive. Also all these classes of positive maps are distinct. Since strong n-positivity means that $\phi \otimes 1_n$ satisfies the generalized Schwarz's inequality, if follows from the above remarks and Theorem 3.1.2 that:

Corollary 3.2.1

Let Δ be a bounded *-linear map on a C^*-algebra A, and define

$$e^{-t\Delta} = \sum_{n=0}^{\infty} \frac{(-t)^n}{n!} \Delta^n \tag{3.2.3}$$

for $t \geq 0$.

The following conditions are equivalent:

1. Δ is a complete dissipation
2. $e^{-t\Delta}$ is a semigroup of completely positive contractions.

We now turn to the classification of complete dissipations. First note that if A does not have an identity, and ϕ is a completely positive map on A, then ϕ has a natural extension to the unit extension of A by setting

$$\phi(\mathbf{1}) = V^*V \tag{3.2.4}$$

where we use the Stinespring decomposition 3.2.1, and then $\phi(\mathbf{1})$ is independent of the decomposition as long as we assume, as we may, that π is non-degenerate. (Actually, the Stinespring decomposition is unique up to unitary equivalence if we assume that $[VH]$ is cyclic for $\pi(A)$).

Theorem 3.2.2 (Christensen-Evans)

Assume that Δ is a bounded *-linear operator on a C^*-algebra A and define $e^{-t\Delta}$ by power series expansion for $t > 0$. Assume that A is represented on a Hilbert space H with σ-weak closure $\bar{A}$.

The following conditions are equivalent

1. $e^{-t\Delta}$ is completely positive for each $t \geq 0$.
2. There exists a completely positive map $\psi : A \to \bar{A}$ and an element $k \in \bar{A}$ such that

$$\Delta(x) = kx + xk^* - \psi(x) \tag{3.2.5}$$

In particular Δ is a complete dissipation if and only if Δ has the decomposition 3.2.4 with

(3.2.6) $$k + k^* - \psi(1) \geq 0$$

Proof

This is Theorem 3.1 in [CE 1]; earlier the result was proved for matrix algebras in [Lin 1]. We follow the proof in [CE 1].

First we show that there exists a Hilbert space K, a representation π of A on K and a linear map $V : A \to L(H,K)$ such that

(3.2.7) $$V(x^*)^*\pi(z)V(y) = \Delta(xz)y + x\Delta(zy) - x\Delta(z)y - \Delta(xzy)$$

for all $x,y,z \in A$,

(3.2.8) $$V(xy) = \pi(x)V(y) + V(x)y$$

for all $x,y \in A$, and

(3.2.9) $$[\pi(A)V(A)H] = K.$$

To this end, let D be the trilinear map on $A \times A \times A$ defined by

$$D(x,z,y) = \Delta(xz)y + x\Delta(zy) - x\Delta(z)y - \Delta(xzy)$$

for $x,z,y \in A$. Then the map $(a_1,a_2) \times (b_1,b_2) \in (A \times A) \times (A \times A) \to D(a_1^*,a_2^*b_2,b_1)$ is positive definite in the sense that the matrix inequality

$$[D(a_{1i}^*,a_{2i}^*a_{2j},a_{1j}]_{i,j} \geq 0$$

is valid for all pairs of finite sequences $a_{11},\dots,a_{1n}$ and $a_{21},\dots,a_{2n}$ in A. By a relatively straightforward Stinespring type argument one now establishes the existence of K, π and V with the properties 3.2.8, 3.2.9 and

$$D(x,z,y) = V(x^*)^*\pi(z)V(y),$$

see Chapter 1 and 2 of [EL 2] or [Lin 2, Propositions 1 and 2] for details.

Next we want to apply Theorem 2.5.11, with the Δ there replaced by V. Note first that 2.5.32 is the same as the relation 3.2.8. By 3.2.9 the identity operator $\mathbf{1}_K$ in $L(K)$ is contained in the weak closure of $\pi(A)$, and letting $\pi(z)$ converge weakly to $\mathbf{1}_K$ in 3.2.7 we deduce that

$$(3.2.10) \qquad V(x^*)^*V(y) \subseteq \bar{A}$$

for all $x,y \in A$, and thus 2.5.31 is fulfilled. Theorem 2.5.11 now implies that there is a $h \in \bar{A}$ such that

$$(3.2.11) \qquad V(x) = hx - \pi(x)h$$

for all $x \in A$. Define the completely positive map ψ from A into $\bar{A}$ by

$$(3.2.12) \qquad \psi(x) = h^*\pi(x)h$$

We have to show there exists a $k \in \bar{A}$ such that

$$(3.2.13) \qquad \Delta(x) + \psi(x) = kx + xk^*.$$

From 3.2.7, 3.2.11 and 3.2.12 it follows that

$$(3.2.14) \qquad \begin{aligned} &\Delta(xz)y + x\Delta(zy) - x\Delta(z)y - \Delta(xzy) \\ &= V(x^*)^*\pi(z)V(y) \\ &= (xh^* - h^*\pi(x))\pi(z)(hy - \pi(y)h) \\ &= \psi(xzy) + x\psi(z)y - x\psi(zy) - \psi(xz)y. \end{aligned}$$

Let e_τ be an approximative unit for A. Then the nets $(\Delta(e_\tau))$ and $(\psi(e_\tau))$ are bounded in $\bar{A}$, and going to a subnet we may therefore assume the nets converge σ-weakly to elements d,p in $\bar{A}$. Put

$$(3.2.15) \qquad a = \tfrac{1}{2}(d+p)$$

By putting $z = e_\tau$ in 3.2.14 and passing to the limit, we obtain

$$(3.2.16) \qquad \begin{aligned} &[\Delta(xy) + \psi(xy)] - x[\Delta(y) + \psi(y)] - [\Delta(x) + \psi(x)]y \\ &= -xdy - xpy = -2xay \\ &= \{a,xy\} - x\{a,y\} - \{a,x\}y \ , \end{aligned}$$

where $\{a,x\} = ax + xy$ denotes the anticommutator. It follows that the map $\delta : A \to \bar{A}$ defined by

$$\delta(x) = \Delta(x) + \psi(x) - \{a,x\} \tag{3.2.17}$$

is a *-derivation. By Sakai-Kadison's derivation theorem there exists a $b = b^* \in \bar{A}$ such that

$$\delta(x) = [ib,x] \tag{3.2.18}$$

Thus the representation 3.2.12 follows, with $k = a+ib$. This ends the proof of $1 \Rightarrow 2$ in Theorem 3.2.2.

To prove $2 \Rightarrow 1$, note that by Arveson's extension theorem, [Arv 1, Theorem 1.2.3], the map ψ extends to a completely positive map from $L(H)$ into $L(H)$, and we denote also the extension of ψ by ψ. The formula 3.2.5 then gives an extension of Δ to $L(H)$ which we also denotes by Δ, and which has a decomposition

$$\Delta = \Delta_1 + \Delta_2 \tag{3.2.19}$$

where

$$\begin{aligned} \Delta_1(x) &= kx + xk^* \\ \Delta_2(x) &= -\psi(x) \end{aligned} \tag{3.2.20}$$

Then $-\Delta_1, -\Delta_2$ are the generators of the semigroups of completely positive maps given by

$$\begin{aligned} e^{-t\Delta_1}(x) &= e^{-tk}x(e^{-tk})^* \\ e^{-t\Delta_2}(x) &= \sum_{n=0}^{\infty} \frac{t^n}{n!}\psi^n(x) \end{aligned} \tag{3.2.21}$$

for $x \in A$, $t \geq 0$. It follows from Trotters product formula

$$e^{-t\Delta}(x) = \lim_{n\to\infty}\left(e^{-\frac{t}{n}\Delta_1} e^{-\frac{t}{n}\Delta_2}\right)^n(x) \tag{3.2.22}$$

that $e^{-t\Delta}$ is a semigroup of completely positive maps on $L(H)$, [BR 1, Corollary 3.1.31], and hence the restriction to A is so. This ends the proof of Theorem 3.2.2.

§3.3 Invariant complete dissipations

We will first give a general description of complete dissipations commuting with a compact abelian action and vanishing on the fixed point algebra for the action. Thereafter we will use the theory of negative definite functions on a group, [BF 1], to give more explisit expressions for the generator in specific cases.

Theorem 3.3.1 (Bratteli-Evans-Jørgensen-Kishimoto-Robinson)

Let G be a compact abelian group, α an action of G on a C^*-algebra A, and $\mathcal{D}$ an α-invariant dense *-subalgebra of A such that $A^\alpha \subseteq \mathcal{D}$ and $\mathcal{D}$ is the linear span of the subspaces $\mathcal{D} \cap A^\alpha(\gamma)$, $\gamma \in \hat{G}$. Furthermore, let Δ be a complete dissipation from $\mathcal{D}$ into A such that

(3.3.1) $\quad \Delta|_{A^\alpha} = 0,$

and

(3.3.2) $\quad [\Delta,\alpha] = 0,$ (See Definition 2.5.9).

It follows that Δ is closable, and its closure $\bar{\Delta}$ generates a strongly continuous semigroup $t \geq 0 \to e^{-t\bar{\Delta}}$ of completely positive contractions.

Furthermore, if A is faithfully and covariantly represented on a Hilbert space H, then for each γ there exists a (possibly unbounded) operator $L(\gamma)$ affiliated with the abelian von Neumann algebra $(M^\alpha \cap (M^\alpha)')E(\gamma)$ such that

(3.3.3) $\quad (\mathcal{D} \cap A^\alpha(\gamma))H \subseteq D(L(\gamma))$

and

(3.3.4) $\quad L(\gamma)x\xi = \Delta(x)\xi$

for each $x \in \mathcal{D} \cap A^\alpha(\gamma)$, $\xi \in H$, where M denotes the σ-weak closure of A in the covariant representation, α denotes also the extension of α to M, and $E(\gamma)$ is the projection onto $[M^\alpha(\gamma)H]$. Let β_γ be the isomorphism from $(M^\alpha \cap (M^\alpha)')E(-\gamma)$ onto $(M^\alpha \cap (M^\alpha)')E(\gamma)$ defined by restricting the β_γ in Lemma 2.7.2. Then each β_γ can be extended to a normal *-morphism of $Z = M^\alpha \cap (M^\alpha)'$ into Z by putting $\beta_\gamma(1-E(-\gamma))$

= 0, and as Z is an abelian von-Neumann algebra, β_γ has also a unique extension to the *-algebra of operators affiliated with Z. If each $L(\gamma)$ is extended to an operator in this algebra by putting $L(\gamma)(\mathbf{1}-E(\gamma)) = 0$, then the map $\gamma \to L(\gamma)$ is a twisted negative-definite function in the sense that the matrix inequality

$$[E(\gamma_i)\{L(\gamma_i)^* + L(\gamma_j) - \beta_{\gamma_i}(L(\gamma_j-\gamma_i))\}E(\gamma_j)] \geq 0 \tag{3.3.5}$$

is valid for all finite sequences $\gamma_1,\dots,\gamma_n$.

Conversely, if $\gamma \to L(\gamma)$ is a twisted negative definite function from $\hat{G}$ into operators affiliated with Z of the above type, and $L(\gamma)(A^\gamma(\gamma) \cap \mathcal{D}) \subseteq A$ for each γ, then the operator Δ defined by 3.3.4 is a complete dissipation with the properties 3.3.1 and 3.3.2.

Proof

A first version of this theorem was proved in [BEv 1] and the present general version was later established in [BJKR 1]. We follow the latter treatment.

First Lemma 2.7.4 implies that

$$\Delta(ax) = a\Delta(x) \ , \quad \Delta(xa) = \Delta(x)a \tag{3.3.6}$$

for all $a \in A^\alpha$, $x \in \mathcal{D}$. The existence of a (possibly unbounded) operator $L(\gamma)$ affiliated with $(M^\alpha \cap (M^\alpha)')E(\gamma)$ such that $(\mathcal{D} \cap A^\alpha(\gamma))H \subseteq D(L(\gamma))$ and $L(\gamma)x = \Delta(x)$ for $x \in \mathcal{D} \cap A^\alpha(\gamma)$ follows exactly as in Lemma 2.7.11. We now argue that $-L(\gamma)$ is dissipative: If $x_i \in A^\alpha(\gamma)$, $\eta_i \in H$ and $\xi = \sum_i x_i\eta_i$, then

$$\begin{aligned} 2\mathrm{Re}(\xi, L(\gamma)\xi) &= (\xi, L(\gamma)\xi) + (L(\gamma)\xi,\xi) \\ &= \sum_{ij} \{(x_i\eta_i, L(\gamma)x_j\eta_j) + (L(\gamma)x_i\eta_i, x_j\eta_j)\} \\ &= \sum_{ij} (\eta_i, \{x_i^*\Delta(x_j) + \Delta(x_i)^*x_j\}\eta_j) \qquad (3.3.7) \\ &\geq \sum_{ij} (\eta_i, \Delta(x_i^*x_j)\eta_j) \\ &= 0 \end{aligned}$$

where we have used that Δ is a complete dissipation and $\Delta|_{A^\alpha} = 0$. We choose $L(\gamma)$ such that $(\mathcal{D} \cap A^\alpha(\gamma))H$ is a core for $L(\gamma)|_{E(\gamma)H}$, and thus it follows that $-L(\gamma)$ is dissipative. Since $L(\gamma)$ is affiliated to the abelian von Neumann algebra $M^\alpha \cap (M^\alpha)'$, this means that the function-representative of $L(\gamma)$ in the spectral representation has non-negative real part, and hence $-L(\gamma)$ generates a semigroup $t > 0 \to e^{-tL(\gamma)}$ of contractions on H.

Now one argues exactly as in Lemma 2.8.3 to show that the set of elements $x \in \mathcal{D} \cap A^\alpha(\gamma)$ such that there exists a $y \in \mathcal{D} \cap A^\alpha(\gamma)$ with $x = yy^*x$ is a norm dense subset of $A^\alpha(\gamma)$, and this subset consists of entire analytic elements for Δ. Exponentiating on this subset and taking limits we see that the semigroup $e^{-tL(\gamma)}$ multiplies $A^\alpha(\gamma)$ into itself to the left, i.e. $e^{-tL(\gamma)} \in$ centre $M(\overline{A^\alpha(\gamma)A^\alpha(\gamma)^{*}}^{\|\,\|})$. Thus we may define a strongly continuous semigroup S of contractions on $A^\alpha(\gamma)$ by

$$(3.3.8) \qquad S_t(x) = e^{-tL(\gamma)}x$$

for $x \in A^\alpha(\gamma)$, $t \geq 0$. We extend S to a semigroup of linear maps from A^α_F into A^α_F by

$$(3.3.9) \qquad S_t(x) = \sum_{\gamma\in\hat{G}} e^{-tL(\gamma)} x_\gamma$$

where $x = \sum_{\gamma\in\hat{G}} x_\gamma$ is the Fourier decomposition of x.

Our aim is now to show that S extends to a semigroup of completely positive contractions, with generator $-\overline{\Delta}$. To this end we first note that all the operators $L(\gamma)$ and $E(\gamma)$, and thus $e^{-tL(\gamma)}$ are contained in $M^\alpha \cap (M^\alpha)'$, and consequently these operators generate an abelian von Neumann algebra. If $\hat{G}$ is countable this algebra is countably decomposable and has the form $L^\infty(\Omega, d\mu)$ for some finite measure μ coming from a normal state with faithful restriction to the abelian algebra. If $\hat{G}$ is not countable one may replace $\hat{G}$ by countable subsets of $\hat{G}$ throughout the following reasoning and assume the existence of a similar representation.

In this spectral representation the projections $E(\gamma)$ are represented by characteristic functions of measurable subsets of Ω, and the operators $L(\gamma)$ by almost everywhere defined measurable functions with nonnegative real part. The morphisms β_γ extend by normality to the latter functions. Then for all $\gamma_1,\dots,\gamma_n \in \hat{G}$ and $m \in \mathbf{N}$ one has

$$(3.3.10) \qquad [E(\gamma_i)\{L(\gamma_i)^* + L(\gamma_j) - \beta_{\gamma_i}(L(\gamma_j - \gamma_i))\}^m E(\gamma_j)] \geq 0$$

i.e. the matrix-valued measurable function from Ω into M_n whose (i,j)-th matrix element is given by the foregoing expression takes values in the positive matrices, almost everywhere. To show this, first note that if $x_i \in D(\gamma_i)$, $i=1,\dots,n$, then

$$(3.3.11) \qquad [\Delta(x_i)^* x_j + x_i^* \Delta(x_j) - \Delta(x_i^* x_j)] \geq 0.$$

Inserting the operators $L(\gamma)$ in this expression, and using Lemma 2.7.2 we obtain

$$(3.3.12) \qquad [x_i^* \{L(\gamma_i)^* + L(\gamma_j) - \beta_{\gamma_i}(L(\gamma_j - \gamma_i))\} x_j] \geq 0$$

By multiplying this to the left and right with the diagonal n×n matrixes with the elements x_i, respectively x_i^*, on the diagonal we obtain

$$(3.3.13) \qquad [x_i x_i^* \{L(\gamma_i)^* + L(\gamma_j) - \beta_{\gamma_i}(L(\gamma_j - \gamma_i))\} x_j x_j^*] \geq 0.$$

Now, by tensoring up the system with the m×m matrixes M_m, and replacing A, S, $L(\gamma)$ and α_g by $A \otimes M_m$, $S \otimes I_m$, $L(\gamma) \otimes 1_m$ and $\alpha_g \otimes \iota_m$, we get a version of 3.3.13 where the x_i's are m×m matrixes with entries in $A^\alpha(\gamma_i)$. Choosing the x_i as a matrix which has the entries $x_i^1, \dots, x_i^m$ in the first row and zeroes elsewhere, and going to submatrises we obtain

$$(3.3.14) \qquad [\sum_k (x_i^k x_i^{k*}) \{L(\gamma_i)^* + L(\gamma_j) - \beta_{\gamma_i}(L(\gamma_j - \gamma_i))\} \sum_k (x_j^k x_j^{k*})] \geq 0$$

for all finite double sequences x_i^k, where $x_i^k \in A^\alpha(\gamma_i)$. But if $\sum_k x_i^k x_i^{k*}$ runs through an approximate identity for $A^\alpha(\gamma_i) A^\alpha(\gamma_i)^*$ (see Lemma 2.5.1), then $\sum_k x_i^k x_i^{k*}$ converges strongly to $E(\gamma_i)$, and hence, working in the spectral representation, one finds

$$(3.3.15) \qquad [E(\gamma_i) \{L(\gamma_i)^* + L(\gamma_j) - \beta_{\gamma_i}(L(\gamma_j - \gamma_i))\} E(\gamma_j)] \geq 0,$$

which is 3.3.10 for $m=1$. But a classical theorem of Schur, [Schu 1], implies that if $[a_{ij}]$ is a positive n×n matrix, then $[(a_{ij})^m]$ is also positive, for $m=1,2,\dots$. Applying Schur's theorem to the matrix valued function in 3.3.15 we finally obtain 3.3.10 for $m=1,2,\dots$.

We will next establish the inequality

$$(3.3.16) \qquad S_t(x^* x) \geq S_t(x)^* S_t(x)$$

for $x \in A_F^\alpha$, $t \geq 0$. To this end, multiply the matrix inequality 3.3.10

to the right with the diagonal $n \times n$ matrix with the entries $e^{-tL(\gamma_i)}$ on the diagonal, and to the left with the adjoint of this matrix, to obtain

$$(3.3.17)\qquad [E(\gamma_i)e^{-tL(\gamma_i)^*}\{L(\gamma_i)^* + L(\gamma_j) - \beta_{\gamma_i}(L(\gamma_j-\gamma_i))\}^m e^{-tL(\gamma_j)}E(\gamma_j)] \geq 0$$

Therefore, if this matrix is multiplied by $t^m/(m!)$, and the sum is taken over $m = 0,1,\dots$, then the sum is larger than the $m = 0$ term, i.e.

$$(3.3.18)\qquad [E(\gamma_i)e^{-t\beta_{\gamma_i}(L(\gamma_j-\gamma_i))}E(\gamma_j)] \geq [E(\gamma_i)e^{-tL(\gamma_i)^*}e^{-tL(\gamma_j)}E(\gamma_j)]$$

and both matrix operators involved in this last inequality have finite norm. Assume that $x \in A_F^\alpha$, and that x has the Fourier decomposition $x = \sum_{i=1}^n x_{\gamma_i} = \sum_{i=1}^n x_i$. Multiply the matrices in 3.3.18 to the right with the $n \times n$ matrix having the entries x_i in the first column and zero elsewhere, and to the left with the adjoint of this matrix. One then obtains matrices where only the upper left hand corner is nonzero, and the resulting inequality becomes, after using $x_i^*E(\gamma_i) = x_i^*$, $E(\gamma_j)x_j = x_j$,

$$(3.3.19)\qquad \sum_{ij} x_i^* e^{-t\beta_{\gamma_i}(L(\gamma_j-\gamma_i))} x_j \geq \sum_{ij} x_i^* e^{-tL(\gamma_i)^*} e^{-tL(\gamma_j)} x_j$$

But

$$(3.3.20)\qquad \begin{aligned} x_i^* e^{-t\beta_{\gamma_i}(L(\gamma_j-\gamma_i))} x_j &= x_i^*\beta_{\gamma_i}(e^{-tL(\gamma_j-\gamma_i)})x_j \\ &= e^{-tL(\gamma_j-\gamma_i)} x_i^* x_j \\ &= S_t(x_i^* x_j) \end{aligned}$$

by 3.3.8 and 2.7.4. Thus the definition 3.3.9 of S_t and the inequality 3.3.19 implies the generalized Schwarz' inequality 3.3.16 for S. Note that by tensoring with the $n \times n$ matrices, one can equally well deduce the matrix inequality

(3.3.21) $[S_t(x_i^* x_j)] \geq [S_t(x_i)^* S_t(x_j)]$

for $x_1,\ldots,x_n \in A_F^\alpha$.

Once 3.3.16 is established, it follows from Lemma 2.5.4 that each S_t is a contraction, and hence S_t extends by continuity to a strongly continuous semigroup of contractions of A. The inequality 3.3.21 extends by continuity to all finite sequences $x_1,\ldots,x_n \in A$, and hence each S_t is a completely positive map.

Now, one uses the argument around 2.8.14-2.8.16 and before Lemma 2.8.4 to show that the generator of S is an extension of Δ, and that $\mathcal{D}$ is a core for this generator. This ends the proof that $-\overline{\Delta}$ is the generator of a semigroup of completely positive contractions. This ends the proof of the main statements of Theorem 3.3.1, and the statements of the last paragraph of the theorem are straightforward.

Note that in the particular case that the centre of M^α is contained in the centre of M, and also that $E(\gamma) = \mathbf{1}$ for all $\gamma \in \hat{G}$, the relation 3.3.5 reduces to

(3.3.22) $[L(\gamma_i)^* + L(\gamma_j) - L(\gamma_j - \gamma_i)] \geq 0$

for all finite sequences $\gamma_1,\ldots,\gamma_n$, i.e. over almost all points in Ω the function $\gamma \to L(\gamma)$ is negative definite, [BF 1]. Combining this with Schoenberg's theorem and Bochner's theorem, one may give an explisit representation of the semi-group $e^{-t\overline{\Delta}}$ in terms of a centre (M^α)-valued measure, see [BEv 1, Theorem 5.1]. We will content ourselves by treating the slightly simpler situation that M^α is a factor:

Corollary 3.3.2 (Bratteli-Evans)

Let G be a compact abelian group, α a faithful action of G on a C^*-algebra A, and assume that there exists an α-covariant faithful representation π of A such that $\pi(A^\alpha)''$ is a factor.

Let $\Delta : D(\Delta) \subseteq A \to A$ be a densely defined, closed, *-linear map.

The following four conditions 1-4 are equivalent

(1) a. $A_F^\alpha \subseteq D(\Delta)$, A_F^α is a core for Δ, and $\Delta|_{A_F^\alpha}$ is a complete dissipation,

b. $[\Delta,\alpha] = 0$,

c. $\Delta|_{A^\alpha} = 0$.

(2) a. $A_F^\alpha \subseteq D(\Delta)$, A_F^α is a core for Δ and there exists a negative definite function λ from $\hat{G}$ into $\mathbb{C}$ such that $\lambda(0) = 0$ and

(3.3.23) $$\Delta(x) = \lambda(\gamma)x$$

for $x \in A^\alpha(\gamma)$, $\gamma \in \hat{G}$.

(λ is said to be negative definite if the matrix inequalities

(3.3.24) $$[\lambda(\gamma_j - \gamma_i) - \overline{\lambda(\gamma_i)} - \lambda(\gamma_j)] \leq 0$$

are valid for all finite sequences $\gamma_1, \dots, \gamma_n \in \hat{G}$.)

(3) $-\Delta$ is the generator of a strongly continuous semigroup $e^{-t\Delta}$ on A such that

a. $e^{-t\Delta}$ is completely positive for each $t \geq 0$.

b. $[e^{-t\Delta}, \alpha] = 0$ for each $t \geq 0$.

c. $e^{-t\Delta}(x) = x$ for all $x \in A^\alpha$, $t \geq 0$.

(4) There exists a convolution semigroup $t \geq 0 \to \mu_t$ of probability measures on G such that $-\Delta$ is the generator of the strongly continuous semigroup S given by

(3.3.25) $$S_t(x) = \int_G d\mu_t(g)\alpha_g(x)$$

for $x \in A$, $t \geq 0$.

(μ_t is said to be a convolution semigroups of measures if $\mu_{t+s} = \mu_t * \mu_s$ for $t, s \geq 0$, $\mu_0 = \delta$ and $t \to \mu_t$ is continous.)

Furthermore, if G is the d-dimensional torus $\mathbf{T}^d$ and $\frac{\partial}{\partial t_i}$, $i = 1, \dots, d$ are the generators of the actions of the canonical one-parameter subgroups of $\mathbf{T}^d$ on A, then conditions (1)-(4) are also equivalent to

(5) $A_F^\alpha \subseteq D(\Delta)$, A_F^α is a core for Δ, and there is a triple (b, a, μ), where $b = (b_1, \dots, b_d)$ is a d-triple of real numbers, $a = [a_{ij}]$ is a real positive $d \times d$ matrix and μ is a non-negative bounded measure on

$$\mathbf{T}^d \setminus \{0\} = \{t = (t_i) \mid -\pi < t_i \leq \pi,\ x \neq 0\}$$

such that

(3.3.26) $$\Delta(x) = \left\{ \sum_{k=1}^d b_k \frac{\partial}{\partial t_k} - \sum_{ij} a_{ij} \frac{\partial^2}{\partial t_i \partial t_j} + \int_{\mathbf{T}^d \setminus \{0\}} \frac{d\mu(t)}{\|t\|^2} \left[1 + \sum_k t_k \frac{\partial}{\partial t_k} - \exp\left(\sum_k t_k \frac{\partial}{\partial t_k} \right) \right] \right\}(x)$$

for $x \in A_F^\alpha$, where $\|t\|^2 = \sum_{k=1} t_k^2$.

Proof

This is Corollary 5.8 in [BEv 1]. The equivalences (1) ⇔ (2) ⇔ (3) are clear from Theorem 3.3.1 and its proof. The equivalence (2) ⇔ (4) is clear from Schoenberg's theorem: $\lambda : \hat{G} \to ¢$ is negative definite if and only if $e^{-t\lambda(\cdot)} : \hat{G} \to ¢$ is positive definite for each $t > 0$, [BF 1, Theorem 8.3] and Bochner's theorem: A function $\phi : \hat{G} \to ¢$ is positive definite with $\phi(0) = 1$ if and only if there exists a probability measure μ on G such that

$$\phi(\gamma) = \int_G d\mu(g)\langle \gamma, g\rangle$$

for $\gamma \in \hat{G}$. The latter measure is unique, so if λ is negative definite with $\lambda(0) = 0$, and μ_t is the probability measure such that

$$e^{-t\lambda(\gamma)} = \int_G d\mu_t(g)\langle \gamma, g\rangle$$

then $t \to \mu_t$ is a convolution semigroup of measures. This gives the equivalence (2) ⇔ (4). Finally (2) ⇔ (5) follows from the Lévy-Khinchin representation of a negative definite function, [PS 1]. See the proof of Corollary 5.8 in [BEv 1] for further details.

3.4. Dissipations vanishing on a fixed point algebra

Throughout Section 3.3 we assumed that the complete dissipation Δ satisfied $\Delta|_{A^\alpha} = 0$ and $[\Delta, \alpha] = 0$. In this section we will consider some cases where $[\Delta, \alpha] = 0$ is not explicitely assumed, but where the dynamical assumptions and $\Delta|_{A^\alpha} = 0$ nevertheless forces the complete dissipation Δ to commute with α.

Theorem 3.4.1

Let G be a compact abelian group, α a faithful action of G on a C^*-algebra A, and assume that there exists a faithful G-covariant representation π of A such that

(3.4.1) $\quad \pi(A^\alpha)' \cap \pi(A)'' = \mathbf{C}1.$

Let $\mathcal{D}$ be an α-invariant dense *-subalgebra of A such that

$A^\alpha \subseteq \mathcal{D}$ and $\mathcal{D}$ is the linear span of the subspaces $\mathcal{D} \cap A^\alpha(\gamma)$, $\gamma \in \hat{G}$. Furthermore, let Δ be a complete dissipation from $\mathcal{D}$ into A such that

$$(3.4.2) \qquad \Delta|_{A^\alpha} = 0$$

It follows that Δ is closable, and its closure $\bar{\Delta}$ generates a strongly continuous semigroup $t \geq 0 \to e^{-t\bar{\Delta}}$ of completely positive contractions on A. Furthermore, there exists a convolution semigroup $t \geq 0 \to \mu_t$ of probability measures on G such that

$$(3.4.3) \qquad e^{-t\bar{\Delta}}(x) = \int_G d\mu_t(g)\alpha_g(x)$$

for all $x \in A$, $t \geq 0$.

Remark 3.4.2 (A. Kishimoto, private communication)

Let G be a compact abelian group and α an action of G on A. If one instead of 3.4.1 assumes that A is simple, A^α is prime and $M(A) \cap (A^\alpha)' = \mathbb{C}1$, then the above theorem remains valid. The crucial step in the proof is that these assumptions imply that $M(B) \cap (B^\alpha)' = \mathbb{C}1$ for all nonzero α-invariant hereditary subalgebras B of A, see [EK 1]. Using this, the proof is almost identical with the proof of Theorem 3.4.1.

Proof of Theorem 3.4.1

Lemma 2.7.4 implies that Δ is a module map over A^α. Suppressing the notation π, the dynamical assumption 3.4.1 implies that the algebras $C(\gamma)$ defined by 2.7.2 all consists of the scalar multiples of the identity only. But then a minor extension of Lemma 2.7.5 implies that $\Delta|_{A^\alpha(\gamma) \cap \mathcal{D}}$ is a scalar multiple of the identity operator for each $\gamma \in \hat{G}$. Thus Δ commutes with α, and Corollary 3.3.2 applies.

The next theorem is similar, but an interesting feature here is that, although Δ is only assumed to be a dissipation, the other assumptions forces Δ to be a complete dissipation. The conditions of this theorem should be compared to those of Theorem 2.9.31.

Theorem 3.4.3 (Kishimoto-Robinson)

Let G be a compact abelian group and α an action of G on a simple unital C^*-algebra A. Assume there exists an automorphism τ of A such that

$$(3.4.4)\qquad [\tau,\alpha] = 0$$

and

$$(3.4.5)\qquad \lim_{|n|\to\infty} \|[x,\tau^n(y)]\| = 0$$

for all $x,y \in A$.

Let $\mathcal{D}$ be an α-invariant dense *-subalgebra of A such that $A^\alpha \subseteq \mathcal{D}$ and $\mathcal{D}$ is the linear span of the subspaces $\mathcal{D} \cap A^\alpha(\gamma)$, $\gamma \in \hat{G}$. Furthermore let $\Delta : \mathcal{D} \to A$ be a dissipation such that

$$(3.4.6)\qquad [\Delta,\tau] = 0$$

and

$$(3.4.7)\qquad \Delta|_{A^\alpha} = 0$$

It follows that Δ is closable, and its closure $\bar{\Delta}$ generates a strongly continuous semigroup $t \geq 0 \to e^{-t\bar{\Delta}}$ of completely positive contractions on A. Furthermore, there exists a convolution semigroup $t \geq 0 \to \mu_t$ of probability measures on G such that

$$(3.4.8)\qquad e^{-t\bar{\Delta}}(x) = \int_G d\mu_t(g)\alpha_g(x)$$

for all $x \in A$, $t \geq 0$.

Proof

This is part of Theorem 3.1 in [KR 3]. We will only outline the argument, and refer to the paper for details:

For each $k = 1,2,\ldots$ there exists a C^*-norm $\|\cdot\|_k$ on the k-fold algebraic tensor product

$$(3.4.9)\qquad A_k = A \odot A \odot \ldots \odot A$$

uniquely defined by

$$(3.4.10)\qquad \|\sum_i x_i^{(1)} \otimes x_i^{(2)} \otimes \dots \otimes x_i^{(k)}\| = \limsup_{|n|\to\infty} \|\sum_i \tau^{n_1}(x_i^{(1)})\tau^{n_2}(x_i^{(2)})\dots\tau^{n_k}(x_i^{(k)})\|$$

where $|n| = \min\{|n_i-n_j| \mid i \neq j\}$. If $x_i \in A^\alpha(\gamma)\cap \mathcal{D}$ are nonzero for $i=1,2,3$ then $\tau^{n_1}(x_1)\tau^{n_2}(x_2^*) \in A^\alpha$ and $\tau^{n_2}(x_2^*)\tau^{n_3}(x_3) \in A^\alpha$, and hence

$$(3.4.11)\qquad \begin{aligned}&\Delta(\tau^{n_1}(x_1)\tau^{n_2}(x_2^*)\tau^{n_3}(x_3))\\ &= \tau^{n_1}(x_1)\tau^{n_2}(x_2^*)\tau^{n_3}(\Delta(x_3))\\ &= \tau^{n_1}(\Delta(x_1))\tau^{n_2}(x_2^*)\tau^{n_3}(x_3)\end{aligned}$$

by Lemma 2.7.4. Taking the limit $|n|\to\infty$ we conclude that

$$(3.4.12)\qquad x_1 \otimes x_2^* \otimes \Delta(x_3) = \Delta(x_1) \otimes x_2^* \otimes x_3$$

and hence

$$(3.4.13)\qquad x_1 \otimes \Delta(x_3) = \Delta(x_1) \otimes x_3$$

for all $x_1, x_3 \in A^\alpha(\gamma)\cap\mathcal{D}$. It immediately follows that there exists a scalar $\lambda(\gamma)$ such that

$$(3.4.14)\qquad \Delta(x) = \lambda(\gamma)x$$

Now one proves that $\gamma \to \lambda(\gamma)$ is a negative definite function in a similar way as in the proof of 3.3.5, but instead of assuming that Δ is a complete dissipation and using the tensor product arguments in the proof of Theorem 3.3.1, it now suffices to assume that Δ is a dissipation and then use the tensor product structure given by 3.4.10. As soon as it has been established that $\lambda(\cdot)$ is negative definite the theorem follows from Corollary 3.3.2.

Chapter 4. Additional remarks

Here we collect some remarks which appeared during the final stages of the preparation of these lecture notes.

Remark 4.1 (A. Ikunishi, private communication)

The estimate 2.3.6 in Lemma 2.3.8 can be improved to

$$(4.1) \qquad \|\hat{\delta}|_{M^\alpha(K)}\| = \|\delta|_{A^\alpha(K)}\|,$$

for all finite subsets $K \subseteq \hat{G}$, and this gives an affirmative answer to Problem 2.3.9. The argument goes as follows: Since $\delta|_{A^\alpha(K)}$ is $\sigma(A^\alpha(K),A^\alpha(K)^*) - \sigma(L(H),L(H)_*)$-continous, δ defines a predual map $\delta_* : L(H)_* \to A^\alpha(K)^*$ with the same norm as δ. Taking the dual of δ_* again we get an extension δ' of δ to a linear map from $\overline{A^\alpha(K)}$ into $L(H)$ with the same norm as δ which is $\sigma(\overline{A^\alpha(K)},A^\alpha(K)^*) - \sigma(L(H),L(H)_*)$ continuous. Here $\overline{A^\alpha(K)}$ is the $\sigma(A^{**},A^*)$-closure of $A^\alpha(K)$ in A^{**}. From this point one argues exactly as in the proof of Theorem 2.3.12, but with δ^{**} there replaced by δ'.

Remark 4.2

In Definition 2.9.6 of condition Γ the condition 2.9.57 can be replaced by the more fundamental requirement:

If $\gamma \in F$ there exists a natural number n and elements $X_1,\ldots,X_n \in A_1^\alpha(\gamma)$ such that $\sum_{k=1}^n X_k^* X_k$ is invertible in $A \otimes M_{d(\gamma)}$.

This is a consequence of the following lemma:

Lemma (Bratteli-Kishimoto-Robinson)

Let α be an action of a compact group G on a unital C^*-algebra A, and let $\gamma \in \hat{G}$.

The following two conditions are equivalent, given $n \in \mathbf{N}$:

1. There exist elements $X_1,\ldots,X_n \in A_1^\alpha(\gamma)$ such that $\sum_{k=1}^n X_k^* X_k$ is invertible.

2. There exists an element $S \in A_n^\alpha(\gamma)$ such that $S^*S = \mathbf{1}_d$.

Proof

$2 \Rightarrow 1$: Let $X_1,\dots,X_n$ be the row-vectors in S, then $X_k \in A_1^\alpha(\gamma)$, $k=1,\dots,n$, and $\sum_{k=1}^{n} X_k^* X_k = S^*S = \mathbf{1}_d$.

$1 \Rightarrow 2$: Put

(4.2) $$T = \begin{pmatrix} X_1 \\ \vdots \\ X_n \end{pmatrix}$$

Then $T \in A_n^\alpha(\gamma)$ and $T^*T = \sum_{k=1}^{n} X_k^* X_k$ is invertible, thus there exists an $\varepsilon > 0$ such that $T^*T \geq \varepsilon \mathbf{1}_d$. If $\delta = \|T^*T\|$, then $\mathrm{Spec}(T^*T) \subseteq [\varepsilon,\delta]$, and hence

(4.3) $$\mathrm{Spec}(TT^*) \subseteq \{0\} \cup [\varepsilon,\delta].$$

Let f be a continuous real function such that

(4.4) $$f(t) = \begin{cases} \delta^{-\frac{1}{2}} & \text{for } t = 0 \\ t^{-\frac{1}{2}} & \text{for } \varepsilon \leq t \leq \delta \end{cases}$$

Define

(4.5) $$A = f(TT^*)$$

Then $A \in A^\alpha \otimes M_n$, and hence

(4.6) $$S \equiv AT \in A_n^\alpha(\gamma).$$

It follows from the definition of A that SS^* is the projection onto the spectral subspace of TT^* corresponding to the spectral interval $[\varepsilon,\delta]$. Thus S^*S is a projection. On the other hand

(4.7) $$S^*S = T^*A^2T \geq \delta^{-1}T^*T \geq \varepsilon\delta^{-1}\mathbf{1}_d,$$

thus the projection S^*S is invertible and

(4.8) $$S^*S = \mathbf{1}_d$$

Remark 4.3 (A. Kishimoto, private communication)

It is remarked in the introduction of §2.9.3, and again before the statement of Theorem 2.9.33 that we have not been able to separate the spatial argument completely from the other arguments in the proof of Theorem 2.9.31 when G is compact and non-abelian. However, the assumptions involving the sequence τ_n in Theorem 2.9.31 may be replaced by the condition that there exists a pure state ω on A with the properties in the conclusion of Theorem 2.9.38, i.e. a pure state such that the center of the direct integral representation

$$(4.9) \qquad \pi = \int_G^{\oplus} dg\ \pi_\omega \circ \alpha_g$$

on $H_\omega \otimes L^2(G)$ is the algebra $\mathbb{1} \otimes L^\infty(G)$. The latter condition is equivalent with the requirement

$$(4.10) \qquad \pi_\omega(A^\alpha)'' = L(H_\omega),$$

see Lemma 3.5 in [EK 1] and Proposition 2.8 in [BEK 1]. But the existence of a faithful irreducible representation π_ω with the property 4.10 suffices to ensure that for any automorphism β of A with $\beta|_{A^\alpha} = \iota$ there exists an $g \in G$ such that $\beta = \alpha_g$, see Theorem 3.4 in [EK 1], or see Remark 2.7 in [BEK 1] for an alternative proof. This may be used in the end of the proof of Theorem 2.9.31 instead of Theorem 2.9.37.

Remark 4.4

Theorem 2.8.16 has recently been extended by A. Ikunishi, [Iku 3], to the case where the condition $\delta|_{A^\alpha} = 0$ is replaced by the weaker condition $A^\alpha \subseteq D(\delta)$, and the new proof simplifies the original proof in [GW 1]. Basically the theory of ergodic action of compact groups on type I von Neumann algebras, [Tak 2], is used to show that δ extends to a generator in the weak closure of any G-covariant representation, and then Lemma 2.8.4 implies that δ is a generator.

Remark 4.5

Let $G = \mathbf{R}$, and α a strongly continuous action of G as a group of isometries on a Banach space B, with C^∞-elements B_∞. We remarked at the end of §2.10 that a dissipiative operator H mapping B_∞ into B_∞ is not necessarily a pregenerator. This is not even true if $B = H$

is a Hilbert space and H is a symmetric operator. Consider for example the harmonic oscillator Hamiltonian $K = p^2 + q^2$ on $\mathcal{H} = L^2(\mathbf{R})$, where $p = -i\frac{d}{dx}$ and $q = x$. Then $\mathcal{H}_\infty = \mathcal{S}(\mathbf{R})$, and any polynomial in p and q maps $\mathcal{H}_\infty$ into $\mathcal{H}_\infty$. But some of these polynomials, e.g. $H = qpq$ and $H = p^2 - q^4$, are symmetric but not essentially self-adjoint. At first sight these examples should give examples of a one-parameter group α on a C^*-algebra $\mathcal{A}$, and a derivation $\delta : \mathcal{A}_\infty \to \mathcal{A}_\infty$ such that δ is not a pregenerator, in two ways:

1. Let $\mathcal{A} = \mathcal{LC}(\mathcal{H})$, $\alpha_t = \mathrm{Ad}(e^{itK})$ and $\delta = \mathrm{ad}(iH)$. But in this case Theorem 2.9.10 implies that any $\delta \in \mathrm{Der}(\mathcal{A}_\infty, \mathcal{A}_\infty)$ (or even $\delta \in \mathrm{Der}(\mathcal{A}_F^\alpha, \mathcal{A})$ provided $\delta|_{\mathcal{A}^\alpha(\Omega)}$ is bounded for a compact neighbourhood Ω of 0 in $\hat{\mathbf{R}} = \mathbf{R}$) is a pregenerator. So $\delta = \mathrm{ad}(iH)$ fails to be defined on $\mathcal{A}_\infty$ even though $H\mathcal{H}_\infty \subseteq \mathcal{H}_\infty$.

2. Let $\mathcal{A}$ be the CAR-algebra over $\mathcal{H}$, α_t the quasi-free group of automorphisms defined by e^{itK}, and δ the quasifree derivation defined by iH, [BR 2], [Bra 4]. Since the Fock vacuum state is pure and α-invariant, it again follows from Theorem 2.9.10 that any $\delta \in \mathrm{Der}(\mathcal{A}_\infty, \mathcal{A}_\infty)$ is a pregenerator. Since the derivation δ defined by iH is obviously not a generator, it again fails to be defined on $\mathcal{A}_F^\alpha$.

Thus it seems very hard to find negative answers to problem 2.1.4, and one may well conjecture that if α is a Lie group action on a C^*-algebra $\mathcal{A}$, then all derivations $\delta \in \mathrm{Der}(\mathcal{A}_\infty, \mathcal{A}_\infty)$ are pregenerators.

References

[AHKT 1] Araki, H., R. Haag, D. Kastler and M. Takesaki, Extension of KMS states and chemical potential, Commun. Math. Phys. 53 (1977), 97-134.

[AP 1] Akemann, C.A. and G.K. Pedersen, Central sequences and inner derivations of separable C*-algebras, Amer. J. Math. 101 (1979), 1047-1061.

[Arv 1] Arveson, W.B., Subalgebras of C*-algebras, Acta Math. 123 (1969), 141-224.

[BaK 1] Batty, C.J.K. and A. Kishimoto, Derivations and one-parameter subgroups of C*-dynamical systems, J. London Math. Soc. 31 (1985), 526-536.

[BaR 1] Batty, C.J.K. and D.W. Robinson, Positive one-parameter semigroups on ordered Banach spaces, Acta Appl. Math. 2 (1984), 221-296.

[BaR 2] Batty, C.J.K. and D.W. Robinson, The characterization of differential operators by locality: Abstract derivations, Erg. Th. Dyn. Syst. 5 (1985), 171-183.

[BaR 3] Batty, C.J.K. and D.W. Robinson, Commutators and generators, Quart. J. Math., submitted.

[Bat 1] Batty, C.J.K., Small perturbations of C*-dynamical systems, Commun. Math. Phys. 68 (1979), 39-43.

[Bat 2] Batty, C.J.K., Unbounded derivations of commutative C*-algebras, Commun. Math. Phys. 61 (1978), 261-266.

[Bat 3] Batty, C.J.K., Connected spaces without derivations, Bull. London Math. Soc. 15 (1983), 349-352.

[Bat 4] Batty, C.J.K., Derivations on compact spaces, Proc. London Math. Soc. (3) 42 (1981), 299-330.

[Bat 5] Batty, C.J.K., Delays to flows on the real line, unpublished (1980).

[Bat 6] Batty, C.J.K., Derivations on the line and flows along orbits, Pacific J. Math., to appear.

[Bat 7] Batty, C.J.K., Local operators and derivations on C*-algebras, Trans. Amer. Math. Soc. 287 (1985), 343-352.

[BCER 1] Batty, C.J.K., A.L. Carey, D.E. Evans and D.W. Robinson, Extending derivations, Publ. RIMS Kyoto Univ. 20 (1984), 119-130.

[BDE 1] Bratteli, O., T. Digernes and G.A. Elliott, Locality and differential operators on C*-algebras II, in H. Araki et. al. eds., Springer LNM 1132 (1985), 46-83.

[BDGR 1] Bratteli, O., T. Digernes, F. Goodman and D.W. Robinson, Integration in abelian C*-dynamical systems, Publ. RIMS Kyoto Univ. 21 (1985), 1001-1030.

[BDR 1] Bratteli, O., T. Digernes and D.W. Robinson, Relative locality of derivations, J. Funct. Anal. 59 (1984), 12-40.

[BEE 1] Bratteli, O., G.A. Elliott and D.E. Evans, Locality and differential operators on C*-algebras, J. Diff. Equations, 64 (1986), 221-273.

[BEGJ 1] Bratteli, O., D.E. Evans, F.M. Goodman and P.E.T. Jørgensen, A dichotomy for derivations on $\mathcal{O}_n$, Publ. RIMS Kyoto Univ. 22 (1986), 103-117.

[BEJ 1] Bratteli, O., G.A. Elliott and P.E.T. Jørgensen, Decomposition of unbounded derivations into invariant and approximately inner parts, J. reine angew. Math. 346 (1984), 166-193.

[BEK 1] Bratteli, O., D.E. Evans and A. Kishimoto, Covariant and extremely non-covariant representations of a C*-algebra, in preparation.

[BEl 1] Bratteli, O. and G.A. Elliott, Structure spaces of approximately finite-dimensional C*-algebras II, J. Funct. Anal. 30 (1978), 74-82.

[BER 1] Bratteli, O., G.A. Elliott and D.W. Robinson, The characterization of differential operators by locality: Classical flows, Compositio Math. 58 (1986), 279-319.

[BER 2] Bratteli, O., G.A. Elliott and D.W. Robinson, Strong topological transitivity and C*-dynamical systems, J. Math. Soc. Japan 37 (1985), 115-133.

[BER 3] Bratteli, O., G.A. Elliott and D.W. Robinson, The characterization of differential operators by locality: Dissipations and ellipticity, Publ. RIMS Kyoto Univ. 21 (1985), 1031-1049.

[BER 4] Bratteli, O., G.A. Elliott and D.W. Robinson, The characterization of differential operators by locality: C*-algebras of type I, J. Operator Theory, to appear.

[BEv 1] Bratteli, O. and D.E. Evans, Dynamical semigroups commuting with compact abelian actions, Ergod. Th. & Dynam. Sys. 3 (1983), 187-217.

[BEv 2] Bratteli, O. and D.E. Evans, Derivations tangential to compact groups: The nonabelian case, Proc. London Math. Soc. 52 (1986), 369-384.

[BF 1] Berg, C. and G. Forst, Potential Theory on Locally Compact Abelian Groups, Springer-Verlag, Berlin-Heidelberg-New York (1975)

[BG 1] Bratteli, O. and F.M. Goodman, Derivations tangential to compact group actions: Spectral conditions in the weak closure, Can. J. Math. 37 (1985), 160-192.

[BGJ 1] Bratteli, O., F.M. Goodman and P.E.T. Jørgensen, Unbounded derivations tangential to compact groups of automorphisms, II, J. Funct. Anal. 61 (1985), 247-289.

[BJ 1] Bratteli, O. and P.E.T. Jørgensen, Unbounded derivations tangential to compact groups of automorphisms, J. Funct. Anal. 48 (1982), 107-133.

[BJ 2] Bratteli, O. and P.E.T. Jørgensen, Derivations commuting with abelian gauge actions on lattice systems, Commun. Math. Phys. 87 (1982), 353-364.

[BJKR 1] Bratteli, O., P.E.T. Jørgensen, A. Kishimoto and D.W. Robinson, A C*-algebraic Schoenberg theorem, Ann. Inst. Fourier 33 (1984), 155-187.

[BK 1] Bratteli, O. and A. Kishimoto, Automatic continuity of derivations on eigen-spaces, Proc. of the Conference on Operator Algebras and Mathematical Physics, Iowa 1985, P.E.T. Jørgensen and P. Muhly eds., Contemporary Mathematics vol. 60, AMS, Providence (1986), to appear.

[BK 2] Bratteli, O. and A. Kishimoto, Derivations and free group actions on C*-algebras, J. Operator Theory 15 (1986), 377-410.

[BKR 1] Bratteli, O., A. Kishimoto and D.W. Robinson, Embedding product type actions into C*-dynamical systems, J. Funct. Anal., to appear.

[Boy 1] Boyadjiev, H.N., Unbounded dissipative operators on Jordan Banach algebras and B*-algebras, Compt. Rend. Acad. Bulgare Sci. 35 (1982), 291-293.

[Boy 2] Boyadziev, H.N., Unbounded generators of positive semigroups on B*-algebras, Compt. Rend. Acad. Bulgare Sci. 35 (1982), 1033-1036.

[BR 1] Bratteli, O. and D.W. Robinson, Operator Algebras and Quantum Statistical Mechanics I, Springer-Verlag, New York-Heidelberg-Berlin (1979)

[BR 2] Bratteli, O. and D.W. Robinson, Operator Algebras and Quantum Statistical Mechanics II, Springer-Verlag, New York- Heidelberg-Berlin (1981)

[BR 3] Bratteli, O. and D.W. Robinson, Positive C_0-semigroups on C*-algebras, Math. Scand. 49 (1981), 259-274.

[BR 4] Bratteli, O. and D.W. Robinson, Unbounded derivations of C*-algebras, Commun. Math. Phys. 42 (1975), 253-268.

[Bra 1] Bratteli, O., Inductive limits of finite-dimensional C*-algebras, Trans. Amer. Math. Soc. 171 (1972), 195-234.

[Bra 2] Bratteli, O., Structure spaces of approximately finite-dimensional C*-algebras, J. Funct. Anal. 16 (1974), 192-204.

[Bra 3] Bratteli, O., On dynamical semigroups and compact group actions, in Springer Lecture Notes in Mathematics 1055, (1984), 46-61.

[Bra 4] Bratteli, O., A remark on extensions of quasifree derivations on the CAR-algebra, Letters Math. Phys. 6 (1982), 499-504.

[Bra 5] Bratteli, O., Derivations and free group actions, Proc. of the Conference on Operator Algebras and Mathematical Physics, Iowa 17-21/6-85, P.E.T. Jørgensen and P. Muhly, eds, Contemporary Mathematics vol. 60, AMS, Providence (1986), to appear.

[Bra 6] Bratteli, O., Unbounded derivations and C*-dynamics, Proc. of the Fifth Symposium of Pure and Applied Mathematics, Kangneung College, Kangwon, Korea 22-27/7-85, eds. Kun Soo Chang and Yong Moon Park, to appear.

[Bro 1] Brown, L.G., Stable isomorphism of hereditary subalgebras of C*-algebras, Pacific J. Math. 71 (1977), 335-348.

[CE 1] Christensen, E. and D.E. Evans, Cohomology of operator algebras and quantum dynamical semigroups, J. London Math. Soc. 20 (1979), 358-368.

[Chi 1] Chi, D.P., Derivations in C*-algebras, Ph.D. thesis, Univ. of Pennsylvania (1976)

[Cho 1] Choi, M.D., Positive linear maps on C*-algebras, Can. J. Math. 24 (1972), 520-529.

[Cho 2] Choi, M.D., Some assorted inequalities for positive linear maps on C*-algebras, J. Operator Theory 4 (1980), 271-285.

[Chr 1] Christensen, E., Extensions of derivations, J. Funct. Anal. 27 (1978), 234-247.

[Chr 2] Christensen, E., Derivations and their relation to perturbation of operator algebras, Proc. Sympos. Pure Math. 38 (1982), 261-274.

[Con 1] Connes, A., Cyclic cohomology and the transverse fundamental class of a foliation, Preprint I.H.E.S. M/84/7 (1984)

[Con 2] Connes, A., Une classification des facteurs de type III, Ann. Sci. Ecole Norm. Sup. 6 (1973), 133-252.

[Con 3] Connes, A., Spectral sequences and homology of currents for operator algebras, Tagungshericht 42/81, Oberwolfach 1981, 4-5.

[Con 4] Connes, A., Non Commutative Differential Geometry, Chapter II: De Rham homology and non commutative algebra, Inst. Hautes Études Sci. Publ. Math. 62 (1985), 257-360.

[Cun 1] Cuntz, J., Simple C*-algebras generated by isometries, Commun. Math. Phys. 57 (1977), 173-185.

[Dav 1] Davies, E.B., A generation theorem for operators commuting with group actions, Math. Proc. Camb. Phil. Soc. 96 (1984), 315-322.

[Dix 1] Dixmier, J., Les algèbres d'opérateurs dans l'espace Hilbertien, 2nd edition Gauthier-Villars, Paris (1969)

[DM 1] Dixmier, J. and P. Malliavin, Factorisations de functions et de vecteurs indefiniment differentiables, Bull. Sci. Math., 2e série, 102 (1978), 305-330.

[DR 1] Doplicher, S. and J.E. Roberts, Compact Lie groups associated with endomorphisms of C*-algebras, Bull. Amer. Math. Soc. 11 (1984), 333-338.

[DS 1] Dunford, N. and J.T. Schwartz, Linear Operators, Part I, Interscience, New York (1967)

[EH 1] Effros, E. and F. Hahn, Locally compact tranformation groups and C*-algebras, Memoirs Amer. Math. Soc. 75 (1967)

[EHO 1] Evans, D.E. and H. Hanche-Olsen, The generators of positive semigroups, J. Funct. Anal. 32 (1979), 207-212.

[EK 1] Evans, D.E. and A. Kishimoto, Duality for automorphisms on a compact C*-dynamical system, preprint (1986).

[EL 1] Evans, D.E. and J.T. Lewis, Dilations of Irreversible Evolutions in Algebraic Quantum Theory, Comm. Dublin Inst. Adv. Studies Series A 24 (1977)

[Ell 1] Elliott, G.A., Some C*-algebras with outer derivations III, Ann. Math. 106 (1977), 121-143.

[Ell 2] Elliott, G.A., Universally weakly inner one-parameter automorphism groups of separable C*-algebras, Math. Scand. 45 (1979), 139-146.

[Eva 1] Evans, D.E., Positive linear maps on operator algebras, Commun. Math. Phys. 48 (1976), 15-22.

[Eva 2] Evans, D.E., Quantum dynamical semigroups, symmetry groups, and locality, Acta Appl. Math. 2 (1984), 333-352.

[Eva 3] Evans, D.E., A review on semigroups of completely positive maps, in K. Osterwalder ed., Mathematical Problems in Theoretical Physics, SLNP116, Springer Verlag, (1980), 400-406.

[GJ 1] Goodman, F.M. and P.E.T. Jørgensen, Unbounded derivations commuting with compact group actions, Commun. Math. Phys. 82 (1981), 399-405.

[GJ 2] Goodman, F.M. and P.E.T. Jørgensen, Lie algebras of unbounded derivations, J. Funct. Anal. 52 (1983), 369-384.

[GJ 3] Goodman, F.M. and P.E.T. Jørgensen, Smooth Lie group actions on non-commutative tori, in preparation.

[GJP 1] Goodman, F.M., P.E.T. Jørgensen and C. Peligrad, Smooth derivations commuting with Lie group actions, Math. Proc. Camb. Phil. Soc. 99 (1986), 307-314.

[Goo 1] Goodman, F.M., Closed derivations in commutative C*-algebras, J. Funct. Anal. 39 (1980), 308-346.

[Goo 2] Goodman, F.M., Translation invariant closed *-derivations, Pacific J. Math. 97 (1981), 403-413.

[Gre 1] Greenleaf, F., Invariant Means on Topological Groups, van Nostrand, New York (1969)

[GW 1] Goodman, F.M. and A.J. Wassermann, Unbounded derivations commuting with compact group actions. II, J. Funct. Anal. 55 (1984), 389-397.

[HR 1] Hewitt, E. and K.A. Ross, Abstract Harmonic Analysis I, Springer Verlag, Berlin-Göttingen-Heidelberg (1963)

[HR 2] Hewitt, E. and K.A. Ross, Abstract Harmonic Analysis II, Springer Verlag, Berlin-Heidelberg-New York (1970)

[Iku 1] Ikunishi, A., Derivations in C*-algebras commuting with compact actions, Publ. RIMS Kyoto Univ. 19 (1983), 99-106.

[Iku 2] Ikunishi, A., Derivations in covariant representations of C*-algebras, Publ. RIMS Kyoto Univ. 22 (1986), in press.

[Irw 1] Irwin, M.C., Smooth Dynamical Systems, Academic Press, London-New York-Toronto-Sydney-San Francisco (1980)

[Joh 1] Johnson, B.E., Continuity of derivations on commutative algebras, Amer. J. Math. 91 (1969), 1-10.

[Jør 1] Jørgensen, P.E.T., New results on unbounded derivations and ergodic groups of automorphisms, Expo. Math. 2 (1984), 3-24.

[Jør 2] Jørgensen, P.E.T., A structure theorem for Lie algebras of unbounded derivations in C*-algebras, Comp. Math. 52 (1984), 85-98.

[Jør 3] Jørgensen, P.E.T., Compact symmetry groups and generators for sub-Markovian semigroups, Z. Wahrscheinlichkeitstheorie verw. Geb. 63 (1983), 1-27.

[Kad 1] Kadison, R.V., editor, Operator Algebras and Applications, Proc. Symp. Pure Math. 38, Part 2, AMS, Providence (1982).

[Kad 2] Kadison, R.V., Derivations on operator algebras, Ann. Math. 83 (1966), 280-293.

[Kad 3] Kadison, R.V., A note on derivations of operator algebras, Bull. London Math. Soc. 7 (1975), 41-44.

[Kap 1] Kaplansky, I., Modules over operator algebras, Amer. J. Math. 75 (1953), 839-859.

[Kis 1] Kishimoto, A., Dissipations and derivations, Commun. Math. Phys. 47 (1976), 25-32.

[Kis 2] Kishimoto, A., Derivations with a domain condition, Yokohama Math. J. 32 (1984), 215-223.

[Kis 3] Kishimoto, A., Automorphisms and covariant irreducible representations, Yokohama Math. J. 31 (1983), 159-168.

[KN 1] Kelley, J.L. and I. Namioka, Linear Topological Spaces, London (1963)

[KR 1] Kishimoto, A. and D.W. Robinson, Derivations, dynamical systems and spectral restrictions, Math. Scand. 56 (1985), 83-95.

[KR 2] Kishimoto, A. and D.W. Robinson, On unbounded derivations commuting with a compact group of *-automorphisms, Publ. RIMS Kyoto Univ. 18 (1982), 1121-1136.

[KR 3] Kishimoto, A. and D.W. Robinson, Dissipations, derivations, dynamical systems, and asymptotic abelianness, J. Operator Theory 13 (1985), 237-253.

[KT 1] Kishimoto, A. and H. Takai, Some remarks on C*-dynamical systems with a compact abelian group. Publ. RIMS Kyoto Univ. 14 (1978), 383-397.

[Kur 1] Kurose, H., An example of a non quasi-wellbehaved derivation in C(I), J. Funct. Anal. 43 (1981), 193-201.

[Kur 2] Kurose, H., Closed derivations in C(I), Tôhoku Math. Journ. 35 (1983), 341-347.

[Kur 3] Kurose, H., On a closed derivation in C(I), Mem. Fac. Sci., Kyushu Univ., Ser A., 36 (1982), 193-198.

[Kur 4] Kurose, H., Closed derivations in C(I), II, unpublished, partly incorporated in [Kur 3].

[Kur 5] Kurose, H., Closed derivations on compact spaces, J. London Math. Soc., to appear.

[Kur 6] Kurose, H., Unbounded *-derivations commuting with actions of $\mathbf{R}^n$ in C*-algebras, Mem. Fac. Sci. Kyushu Univ. Ser. A 37 (1983), 107-112.

[Lin 1] Lindblad, G., On the generators of quantum dynamical semigroups, Commun. Math. Phys. 48 (1976), 119-130.

[Lin 2] Lindblad, G., Dissipative operators and cohomology of operator algebras, Letters in Math. Phys. 1 (1976), 219-224.

[LN 1] Lance, C. and A. Niknam, Unbounded derivations of group C*-algebras, Proc. Amer. Math. Soc. 61 (1976), 310-314.

[Lon 1] Longo, R., Automatic relative boundedness of derivations in C*-algebras, J. Funct. Anal. 34 (1979), 21-28.

[LP 1] Longo, R. and C. Peligrad, Nón commutative topological dynamics and compact actions on C*-algebras, J. Funct. Anal. 58 (1984), 157-174.

[LTW 1] Lazar, A.J., S.K. Tsui and S. Wright, Derivations from subalgebras of separable C*-algebras, Michigan Math. J. 31 (1984), 65-72.

[LTW 2] Lazar, A.J., S.K. Tsui and S. Wright, A cohomological characterization of finite-dimensional C*-algebras, J. Operator Theory 14 (1985), 239-247.

[Maz 1] Mazur, H., Interval exchange transformations and measured foliations, Ann. Math. 115 (1982), 169-200.

[McI 1] Mc Intosh, A., Functions and derivations of C*-algebras, J. Funct. Anal. 30 (1978), 264-275.

[Nak 1] Nakazato, H., Closed *-derivations on compact groups, J. Math. Soc. Japan 34 (1982), 83-93.

[Nak 2] Nakazato, H., On left invariant dissipative operators, Arch. Math. 45 (1985), 458-462.

[Nak 3] Nakazato, H., Extension of derivations in the algebra of compact operators, J. Funct. Anal. 57 (1984), 101-110.

[Nik 1] Niknam, A., Closable derivations of simple C*-algebras, Glasgow Math. J. 24 (1983), 181-183.

[Nis 1] Nishio, K., A local kernel property of closed derivations on C(I×I), Proc. Amer. Math. Soc. 95 (1985), 573-576.

[OP 1] Olesen, D. and G.K. Pedersen, Applications of the Connes spectrum to C*-dynamical systems III, J. Funct. Anal. 45 (1982), 357-390.

[OPT 1] Olesen, D., G.K. Pedersen and M. Takesaki, Ergodic actions of compact abelian groups, J. Operator Theory 3 (1980) 237-269.

[Ped 1] Pedersen, G.K., C*-algebras and their Automorphism Groups, Academic Press, London-New York-San Francisco (1979)

[Pel 1] Peligrad, C., Derivations of C*-algebras which are invariant under an automorphism group, in "Topics in Modern Operator Theory", OT Series vol. 2, Birkhäuser Verlag (1981), 259-268.

[Pel 2] Peligrad, C., Derivations of C*-algebras which are invariant under an automorphism group. II, in Invariant subspaces and other topics (Timisoara/Herculane, 1981), Operator Theory: Adv. Appl. 6, Birkhäuser, Basel-Boston (1982).

[Pou 1] Poulsen, N.S., On C^∞-vectors and intertwining bilinear forms for representations of Lie groups, J. Funct. Anal. 9 (1972), 87-120.

[PP 1] Powers, R.T. and G. Price, Derivations vanishing on $S(\infty)$, Commun. Math. Phys. 84 (1982), 439-447.

[Pri 1] Price, G.L., Extensions of quasi-free derivations on the CAR algebra, Publ. RIMS Kyoto Univ. 19 (1983), 345-354.

[Pri 2] Price, G.L., On some non-extendable derivations of the gauge-invariant CAR algebra, Trans. Amer. Math. Soc. 285 (1984), 185-201.

[PS 1] Parthasarathy, K.R. and K. Schmidt, Positive Definite Kernels, Continuous Tensor Products, and Central Limit Theorems of Probability Theory, SLM 272, Springer-Verlag, Berlin-Heidelberg-New York (1972).

[Rin 1] Ringrose, J.R., Automatic continuity of derivations of operator algebras, J. London Math. Soc. 5 (1972), 432-438.

[Rob 1] Robinson, D.W., Smooth derivations on abelian C*-dynamical systems, J. Austral. Math. Soc., Series A, to appear.

[Rob 2] Robinson, D.W., Commutators and generators II, Quart. J. Math., submitted.

[Rob 3] Robinson, D.W., Smooth cores of Lipschitz flows, Publ. RIMS Kyoto Univ., to appear.

[Rob 4] Robinson, D.W., Differential operators on C*-algebras, in "Operator Algebras and Mathematical Physics", eds. P.E.T. Jørgensen and P. Muhly, Contemporary Mathematics vol. 60, AMS, Providence (1986).

[Rot 1] Roth, J.P., Opérateurs dissipatifs et semigroupes dans les espaces de fonctions continues, Ann. Inst. Fourier 26 (1976), 1-97.

[RST 1] Robinson, D.W., E. Størmer and M. Takesaki, Derivations of simple C*-algebras tangential to compact automorphism groups, J. Operator Theory 13 (1985), 189-200.

[Rud 1] Rudin, W., Fourier Analysis on Groups, Interscience Publishers (1967)

[Sak 1] Sakai, S., On a conjecture of Kaplansky, Tôhoku Math. Journ. 12 (1960), 31-33.

[Sak 2] Sakai, S., Derivations of W*-algebras, Ann. Math. 83 (1966), 273-279.

[Sak 3] Sakai, S., Derivations of simple C*-algebras, J. Funct. Anal. 2 (1968), 202-206.

[Sak 4] Sakai, S., On one parameter groups of *-automorphisms on operator algebras and the corresponding unbounded derivations, Amer. J. Math. 98 (1976), 427-440.

[Sak 5] Sakai, S., Theory of unbounded derivations in C*-algebras, Lecture notes, Copenhagen Univ. and Univ. of Newcastle upon Tyre (1977).

[Sak 6] Sakai, S., C*-algebras and W*-algebras, Springer-Verlag, Berlin-Heidelberg-New York (1971).

[Sch 1] Schmidt, W.M., Diophantine Approximation, Springer Lecture Notes in Mathematics 785 (1980).

[Schu 1] Schur, I., Bemerkungen zur theorie der beschränkten bilinearformen mit unendlich vielen veränderlichen, J. Reine Ang. Math. 140 (1911), 1-28.

[Seg 1] Segal, I.E., A non-commutative extension of abstract integration, Ann. Math. 57 (1953), 401-457.

[Sti 1] Stinespring, W.F., Positive functions on C*-algebras, Proc. Amer. Math. Soc. 6 (1955), 211-216.

[Tak 1] Takesaki, M., Fourier analysis of compact automorphism groups (An application of Tannaka duality theorem), in D. Kastler ed., Algèbres d'opérateurs et leurs applications en physique mathématique, Éditions CNRS, Paris (1979).

[Taka 1] Takai, H., On a problem of Sakai in unbounded derivations, J. Funct. Anal. 43 (1981), 202-208.

[Tho 1] Thomsen, K., A note to the previous paper by Bratteli, Goodman and Jørgensen, J. Funct. Anal. 61 (1985), 290-294.

[Tho 2] Thomsen, K., Dispersiveness and positive contractive semigroups, J. Funct. Anal. 56 (1984), 348-359.

[Tom 1] Tomiyama, J., The Theory of Closed Derivations in the Algebra of Continous Functions on the Unit Interval, Lecture notes, Tsing Hua Univ. (1983).

[Tom 2] Tomiyama, J., On the closed derivations on the unit interval, J. Ramanujan Math. Soc., submitted.

[Vee 1] Veech, W.A., Gauss measures for transformations on the space of interval exchange maps, Ann. Math. 115 (1982), 201-242.

[Was 1] Wassermann, A.J., Automorphic actions of compact groups on operator algebras, Univ. of Pennsylvania Thesis, 1981.

[Iku 3] Ikunishi, A, The W^*-dynamical system associated with a C^*-dynamical system, and unbounded derivations, Senshu preprint (1986).

[JP 1] Jørgensen, P.E.T. and G.L. Price, Extending quasi-free derivations on the CAR-algebra, J. Operator Theory, to appear.

[LTW 3] Lazar, A.J., S.K. Tsui and S. Wright, Derivations from hereditary subalgebras of C^*-algebras, Canadian J. Math., to appear.

[LTW 4] Lazar, A.J., S.K. Tsui and S. Wright, Extending derivations from hereditary subalgebras of approximately finite-dimensional C^*-algebras, J. London Math. Soc., to appear.

[Tak 2] Takesaki, M., Duality for crossed products and the structure of von Neumann algebras of type III, Acta Math. 131 (1973), 249-310.

[Pow 1] Powers, R.T., An index theory for semigroups of *-endomorphisms of $\mathcal{B}(\mathcal{H})$ and type II_1 factors, Pennsylvania preprint (1986)

SUBJECT INDEX

Vol. 1062: J. Jost, Harmonic Maps Between Surfaces. X, 133 pages. 1984.

Vol. 1063: Orienting Polymers. Proceedings, 1983. Edited by J.L. Ericksen. VII, 166 pages. 1984.

Vol. 1064: Probability Measures on Groups VII. Proceedings, 1983. Edited by H. Heyer. X, 588 pages. 1984.

Vol. 1065: A. Cuyt, Padé Approximants for Operators: Theory and Applications. IX, 138 pages. 1984.

Vol. 1066: Numerical Analysis. Proceedings, 1983. Edited by D.F. Griffiths. XI, 275 pages. 1984.

Vol. 1067: Yasuo Okuyama, Absolute Summability of Fourier Series and Orthogonal Series. VI, 118 pages. 1984.

Vol. 1068: Number Theory, Noordwijkerhout 1983. Proceedings. Edited by H. Jager. V, 296 pages. 1984.

Vol. 1069: M. Kreck, Bordism of Diffeomorphisms and Related Topics. III, 144 pages. 1984.

Vol. 1070: Interpolation Spaces and Allied Topics in Analysis. Proceedings, 1983. Edited by M. Cwikel and J. Peetre. III, 239 pages. 1984.

Vol. 1071: Padé Approximation and its Applications, Bad Honnef 1983. Prodeedings. Edited by H. Werner and H.J. Bünger. VI, 264 pages. 1984.

Vol. 1072: F. Rothe, Global Solutions of Reaction-Diffusion Systems. V, 216 pages. 1984.

Vol. 1073: Graph Theory, Singapore 1983. Proceedings. Edited by K.M. Koh and H.P. Yap. XIII, 335 pages. 1984.

Vol. 1074: E.W. Stredulinsky, Weighted Inequalities and Degenerate Elliptic Partial Differential Equations. III, 143 pages. 1984.

Vol. 1075: H. Majima, Asymptotic Analysis for Integrable Connections with Irregular Singular Points. IX, 159 pages. 1984.

Vol. 1076: Infinite-Dimensional Systems. Proceedings, 1983. Edited by F. Kappel and W. Schappacher. VII, 278 pages. 1984.

Vol. 1077: Lie Group Representations III. Proceedings, 1982–1983. Edited by R. Herb, R. Johnson, R. Lipsman, J. Rosenberg. XI, 454 pages. 1984.

Vol. 1078: A.J.E.M. Janssen, P. van der Steen, Integration Theory. V, 224 pages. 1984.

Vol. 1079: W. Ruppert. Compact Semitopological Semigroups: An Intrinsic Theory. V, 260 pages. 1984

Vol. 1080: Probability Theory on Vector Spaces III. Proceedings, 1983. Edited by D. Szynal and A. Weron. V, 373 pages. 1984.

Vol. 1081: D. Benson, Modular Representation Theory: New Trends and Methods. XI, 231 pages. 1984.

Vol. 1082: C.-G. Schmidt, Arithmetik Abelscher Varietäten mit komplexer Multiplikation. X, 96 Seiten. 1984.

Vol. 1083: D. Bump, Automorphic Forms on GL (3,IR). XI, 184 pages. 1984.

Vol. 1084: D. Kletzing, Structure and Representations of Q-Groups. VI, 290 pages. 1984.

Vol. 1085: G.K. Immink, Asymptotics of Analytic Difference Equations. V, 134 pages. 1984.

Vol. 1086: Sensitivity of Functionals with Applications to Engineering Sciences. Proceedings, 1983. Edited by V. Komkov. V, 130 pages. 1984

Vol. 1087: W. Narkiewicz, Uniform Distribution of Sequences of Integers in Residue Classes. VIII, 125 pages. 1984.

Vol. 1088: A.V. Kakosyan, L.B. Klebanov, J.A. Melamed, Characterization of Distributions by the Method of Intensively Monotone Operators. X, 175 pages. 1984.

Vol. 1089: Measure Theory, Oberwolfach 1983. Proceedings. Edited by D. Kölzow and D. Maharam-Stone. XIII, 327 pages. 1984.

Vol. 1090: Differential Geometry of Submanifolds. Proceedings, 1984. Edited by K. Kenmotsu. VI, 132 pages. 1984.

Vol. 1091: Multifunctions and Integrands. Proceedings, 1983. Edited by G. Salinetti. V, 234 pages. 1984.

Vol. 1092: Complete Intersections. Seminar, 1983. Edited by S. Greco and R. Strano. VII, 299 pages. 1984.

Vol. 1093: A. Prestel, Lectures on Formally Real Fields. XI, 125 pages. 1984.

Vol. 1094: Analyse Complexe. Proceedings, 1983. Edité par E. Amar, R. Gay et Nguyen Thanh Van. IX, 184 pages. 1984.

Vol. 1095: Stochastic Analysis and Applications. Proceedings, 1983. Edited by A. Truman and D. Williams. V, 199 pages. 1984.

Vol. 1096: Théorie du Potentiel. Proceedings, 1983. Edité par G. Mokobodzki et D. Pinchon. IX, 601 pages. 1984.

Vol. 1097: R.M. Dudley, H. Kunita, F. Ledrappier, École d'Éte de Probabilités de Saint-Flour XII – 1982. Edité par P.L. Hennequin. X, 396 pages. 1984.

Vol. 1098: Groups – Korea 1983. Proceedings. Edited by A.C. Kim and B.H. Neumann. VII, 183 pages. 1984.

Vol. 1099: C.M. Ringel, Tame Algebras and Integral Quadratic Forms. XIII, 376 pages. 1984.

Vol. 1100: V. Ivrii, Precise Spectral Asymptotics for Elliptic Operators Acting in Fiberings over Manifolds with Boundary. V, 237 pages. 1984.

Vol. 1101: V. Cossart, J. Giraud, U. Orbanz, Resolution of Surface Singularities. Seminar. VII, 132 pages. 1984.

Vol. 1102: A. Verona, Stratified Mappings – Structure and Triangulability. IX, 160 pages. 1984.

Vol. 1103: Models and Sets. Proceedings, Logic Colloquium, 1983, Part I. Edited by G.H. Müller and M.M. Richter. VIII, 484 pages. 1984.

Vol. 1104: Computation and Proof Theory. Proceedings, Logic Colloquium, 1983, Part II. Edited by M.M. Richter, E. Börger, W. Oberschelp, B. Schinzel and W. Thomas. VIII, 475 pages. 1984.

Vol. 1105: Rational Approximation and Interpolation. Proceedings, 1983. Edited by P.R. Graves-Morris, E.B. Saff and R.S. Varga. XII, 528 pages. 1984.

Vol. 1106: C.T. Chong, Techniques of Admissible Recursion Theory. IX, 214 pages. 1984.

Vol. 1107: Nonlinear Analysis and Optimization. Proceedings, 1982. Edited by C. Vinti. V, 224 pages. 1984.

Vol. 1108: Global Analysis – Studies and Applications I. Edited by Yu.G. Borisovich and Yu.E. Gliklikh. V, 301 pages. 1984.

Vol. 1109: Stochastic Aspects of Classical and Quantum Systems. Proceedings, 1983. Edited by S. Albeverio, P. Combe and M. Sirugue-Collin. IX, 227 pages. 1985.

Vol. 1110: R. Jajte, Strong Limit Theorems in Non-Commutative Probability. VI, 152 pages. 1985.

Vol. 1111: Arbeitstagung Bonn 1984. Proceedings. Edited by F. Hirzebruch, J. Schwermer and S. Suter. V, 481 pages. 1985.

Vol. 1112: Products of Conjugacy Classes in Groups. Edited by Z. Arad and M. Herzog. V, 244 pages. 1985.

Vol. 1113: P. Antosik, C. Swartz, Matrix Methods in Analysis. IV, 114 pages. 1985.

Vol. 1114: Zahlentheoretische Analysis. Seminar. Herausgegeben von E. Hlawka. V, 157 Seiten. 1985.

Vol. 1115: J. Moulin Ollagnier, Ergodic Theory and Statistical Mechanics. VI, 147 pages. 1985.

Vol. 1116: S. Stolz, Hochzusammenhängende Mannigfaltigkeiten und ihre Ränder. XXIII, 134 Seiten. 1985.

Vol. 1117: D.J. Aldous, J.A. Ibragimov, J. Jacod, Ecole d'Été de Probabilités de Saint-Flour XIII – 1983. Édité par P.L. Hennequin. IX, 409 pages. 1985.

Vol. 1118: Grossissements de filtrations: exemples et applications. Seminaire, 1982/83. Edité par Th. Jeulin et M. Yor. V, 315 pages. 1985.

Vol. 1119: Recent Mathematical Methods in Dynamic Programming. Proceedings, 1984. Edited by I. Capuzzo Dolcetta, W.H. Fleming and T. Zolezzi. VI, 202 pages. 1985.

Vol. 1120: K. Jarosz, Perturbations of Banach Algebras. V, 118 pages. 1985.

Vol. 1121: Singularities and Constructive Methods for Their Treatment. Proceedings, 1983. Edited by P. Grisvard, W. Wendland and J.R. Whiteman. IX, 346 pages. 1985.

Vol. 1122: Number Theory. Proceedings, 1984. Edited by K. Alladi. VII, 217 pages. 1985.

Vol. 1123: Séminaire de Probabilités XIX 1983/84. Proceedings. Edité par J. Azéma et M. Yor. IV, 504 pages. 1985.

Vol. 1124: Algebraic Geometry, Sitges (Barcelona) 1983. Proceedings. Edited by E. Casas-Alvero, G.E. Welters and S. Xambó-Descamps. XI, 416 pages. 1985.

Vol. 1125: Dynamical Systems and Bifurcations. Proceedings, 1984. Edited by B.L.J. Braaksma, H.W. Broer and F. Takens. V, 129 pages. 1985.

Vol. 1126: Algebraic and Geometric Topology. Proceedings, 1983. Edited by A. Ranicki, N. Levitt and F. Quinn. V, 423 pages. 1985.

Vol. 1127: Numerical Methods in Fluid Dynamics. Seminar. Edited by F. Brezzi, VII, 333 pages. 1985.

Vol. 1128: J. Elschner, Singular Ordinary Differential Operators and Pseudodifferential Equations. 200 pages. 1985.

Vol. 1129: Numerical Analysis, Lancaster 1984. Proceedings. Edited by P.R. Turner. XIV, 179 pages. 1985.

Vol. 1130: Methods in Mathematical Logic. Proceedings, 1983. Edited by C.A. Di Prisco. VII, 407 pages. 1985.

Vol. 1131: K. Sundaresan, S. Swaminathan, Geometry and Nonlinear Analysis in Banach Spaces. III, 116 pages. 1985.

Vol. 1132: Operator Algebras and their Connections with Topology and Ergodic Theory. Proceedings, 1983. Edited by H. Araki, C.C. Moore, Ş. Strătilă and C. Voiculescu. VI, 594 pages. 1985.

Vol. 1133: K.C. Kiwiel, Methods of Descent for Nondifferentiable Optimization. VI, 362 pages. 1985.

Vol. 1134: G.P. Galdi, S. Rionero, Weighted Energy Methods in Fluid Dynamics and Elasticity. VII, 126 pages. 1985.

Vol. 1135: Number Theory, New York 1983–84. Seminar. Edited by D.V. Chudnovsky, G.V. Chudnovsky, H. Cohn and M.B. Nathanson. V, 283 pages. 1985.

Vol. 1136: Quantum Probability and Applications II. Proceedings, 1984. Edited by L. Accardi and W. von Waldenfels. VI, 534 pages. 1985.

Vol. 1137: Xiao G., Surfaces fibrées en courbes de genre deux. IX, 103 pages. 1985.

Vol. 1138: A. Ocneanu, Actions of Discrete Amenable Groups on von Neumann Algebras. V, 115 pages. 1985.

Vol. 1139: Differential Geometric Methods in Mathematical Physics. Proceedings, 1983. Edited by H. D. Doebner and J. D. Hennig. VI, 337 pages. 1985.

Vol. 1140: S. Donkin, Rational Representations of Algebraic Groups. VII, 254 pages. 1985.

Vol. 1141: Recursion Theory Week. Proceedings, 1984. Edited by H.-D. Ebbinghaus, G.H. Müller and G.E. Sacks. IX, 418 pages. 1985.

Vol. 1142: Orders and their Applications. Proceedings, 1984. Edited by I. Reiner and K. W. Roggenkamp. X, 306 pages. 1985.

Vol. 1143: A. Krieg, Modular Forms on Half-Spaces of Quaternions. XIII, 203 pages. 1985.

Vol. 1144: Knot Theory and Manifolds. Proceedings, 1983. Edited by D. Rolfsen. V, 163 pages. 1985.

Vol. 1145: G. Winkler, Choquet Order and Simplices. VI, 143 pages. 1985.

Vol. 1146: Séminaire d'Algèbre Paul Dubreil et Marie-Paule Malliavin. Proceedings, 1983–1984. Edité par M.-P. Malliavin. IV, 420 pages. 1985.

Vol. 1147: M. Wschebor, Surfaces Aléatoires. VII, 111 pages. 1985.

Vol. 1148: Mark A. Kon, Probability Distributions in Quantum Statistical Mechanics. V, 121 pages. 1985.

Vol. 1149: Universal Algebra and Lattice Theory. Proceedings, 1984. Edited by S. D. Comer. VI, 282 pages. 1985.

Vol. 1150: B. Kawohl, Rearrangements and Convexity of Level Sets in PDE. V, 136 pages. 1985.

Vol 1151: Ordinary and Partial Differential Equations. Proceedings, 1984. Edited by B.D. Sleeman and R.J. Jarvis. XIV, 357 pages. 1985.

Vol. 1152: H. Widom, Asymptotic Expansions for Pseudodifferential Operators on Bounded Domains. V, 150 pages. 1985.

Vol. 1153: Probability in Banach Spaces V. Proceedings, 1984. Edited by A. Beck, R. Dudley, M. Hahn, J. Kuelbs and M. Marcus. VI, 457 pages. 1985.

Vol. 1154: D.S. Naidu, A.K. Rao, Singular Pertubation Analysis of Discrete Control Systems. IX, 195 pages. 1985.

Vol. 1155: Stability Problems for Stochastic Models. Proceedings, 1984. Edited by V.V. Kalashnikov and V.M. Zolotarev. VI, 447 pages. 1985.

Vol. 1156: Global Differential Geometry and Global Analysis 1984. Proceedings, 1984. Edited by D. Ferus, R.B. Gardner, S. Helgason and U. Simon. V, 339 pages. 1985.

Vol. 1157: H. Levine, Classifying Immersions into $\mathbb{R}^4$ over Stable Maps of 3-Manifolds into $\mathbb{R}^2$. V, 163 pages. 1985.

Vol. 1158: Stochastic Processes – Mathematics and Physics. Proceedings, 1984. Edited by S. Albeverio, Ph. Blanchard and L. Streit. VI, 230 pages. 1986.

Vol. 1159: Schrödinger Operators, Como 1984. Seminar. Edited by S. Graffi. VIII, 272 pages. 1986.

Vol. 1160: J.-C. van der Meer, The Hamiltonian Hopf Bifurcation. VI, 115 pages. 1985.

Vol. 1161: Harmonic Mappings and Minimal Immersions, Montecatini 1984. Seminar. Edited by E. Giusti. VII, 285 pages. 1985.

Vol. 1162: S.J.L. van Eijndhoven, J. de Graaf, Trajectory Spaces, Generalized Functions and Unbounded Operators. IV, 272 pages. 1985.

Vol. 1163: Iteration Theory and its Functional Equations. Proceedings, 1984. Edited by R. Liedl, L. Reich and Gy. Targonski. VIII, 231 pages. 1985.

Vol. 1164: M. Meschiari, J.H. Rawnsley, S. Salamon, Geometry Seminar "Luigi Bianchi" II – 1984. Edited by E. Vesentini. VI, 224 pages. 1985.

Vol. 1165: Seminar on Deformations. Proceedings, 1982/84. Edited by J. Ławrynowicz. IX, 331 pages. 1985.

Vol. 1166: Banach Spaces. Proceedings, 1984. Edited by N. Kalton and E. Saab. VI, 199 pages. 1985.

Vol. 1167: Geometry and Topology. Proceedings, 1983–84. Edited by J. Alexander and J. Harer. VI, 292 pages. 1985.

Vol. 1168: S.S. Agaian, Hadamard Matrices and their Applications. III, 227 pages. 1985.

Vol. 1169: W.A. Light, E.W. Cheney, Approximation Theory in Tensor Product Spaces. VII, 157 pages. 1985.

Vol. 1170: B.S. Thomson, Real Functions. VII, 229 pages. 1985.

Vol. 1171: Polynômes Orthogonaux et Applications. Proceedings, 1984. Edité par C. Brezinski, A. Draux, A.P. Magnus, P. Maroni et A. Ronveaux. XXXVII, 584 pages. 1985.

Vol. 1172: Algebraic Topology, Göttingen 1984. Proceedings. Edited by L. Smith. VI, 209 pages. 1985.